PROGRESS IN CLINICAL AND BIOLOGICAL RESEARCH

Series Editors

Nathan Back   Vincent P. Eijsvoogel   Kurt Hirschhorn   Sidney Udenfriend
George J. Brewer   Robert Grover   Seymour S. Kety   Jonathan W. Uhr

RECENT TITLES

Vol 237: **The Use of Transrectal Ultrasound in the Diagnosis and Management of Prostate Cancer,** Fred Lee, Richard McLeary, *Editors*

Vol 238: **Avian Immunology,** W.T. Weber, D.L. Ewert, *Editors*

Vol 239: **Current Concepts and Approaches to the Study of Prostate Cancer,** Donald S. Coffey, Nicholas Bruchovsky, William A. Gardner, Jr., Martin I. Resnick, James P. Karr, *Editors*

Vol 240: **Pathophysiological Aspects of Sickle Cell Vaso-Occlusion,** Ronald L. Nagel, *Editor*

Vol 241: **Genetics and Alcoholism,** H. Werner Goedde, Dharam P. Agarwal, *Editors*

Vol 242: **Prostaglandins in Clinical Research,** Helmut Sinzinger, Karsten Schrör, *Editors*

Vol 243: **Prostate Cancer,** Gerald P. Murphy, Saad Khoury, Réne Küss, Christian Chatelain, Louis Denis, *Editors*. Published in two volumes: Part A: *Research, Endocrine Treatment, and Histopathology*. Part B: *Imaging Techniques, Radiotherapy, Chemotherapy, and Management Issues*

Vol 244: **Cellular Immunotherapy of Cancer,** Robert L. Truitt, Robert P. Gale, Mortimer M. Bortin, *Editors*

Vol 245: **Regulation and Contraction of Smooth Muscle,** Marion J. Siegman, Andrew P. Somlyo, Newman L. Stephens, *Editors*

Vol 246: **Oncology and Immunology of Down Syndrome,** Ernest E. McCoy, Charles J. Epstein, *Editors*

Vol 247: **Degenerative Retinal Disorders: Clinical and Laboratory Investigations,** Joe G. Hollyfield, Robert E. Anderson, Matthew M.LaVail, *Editors*

Vol 248: **Advances in Cancer Control: The War on Cancer—15 Years of Progress,** Paul F. Engstrom, Lee E. Mortenson, Paul N. Anderson, *Editors*

Vol 249: **Mechanisms of Signal Transduction by Hormones and Growth Factors,** Myles C. Cabot, Wallace L. McKeehan, *Editors*

Vol 250: **Kawasaki Disease,** Stanford T. Shulman, *Editor*

Vol 251: **Developmental Control of Globin Gene Expression,** George Stamatoyannopoulos, Arthur W. Nienhuis, *Editors*

Vol 252: **Cellular Calcium and Phosphate Transport in Health and Disease,** Felix Bronner, Meinrad Peterlik, *Editors*

Vol 253: **Model Systems in Neurotoxicology: Alternative Approaches to Animal Testing,** Abraham Shahar, Alan M. Goldberg, *Editors*

Vol 254: **Genetics and Epithelial Cell Dysfunction in Cystic Fibrosis,** John R. Riordan, Manuel Buchwald, *Editors*

Vol 255: **Recent Aspects of Diagnosis and Treatment of Lipoprotein Disorders: Impact on Prevention of Atherosclerotic Diseases,** Kurt Widhalm, Herbert K. Naito, *Editors*

Vol 256: **Advances in Pigment Cell Research,** Joseph T. Bagnara, *Editor*

Vol 257: **Electromagnetic Fields and Neurobehavioral Function,** Mary Ellen O'Connor, Richard H. Lovely, *Editors*

Vol 258: **Membrane Biophysics III: Biological Transport,** Mumtaz A. Dinno, William McD. Armstrong, *Editors*

Vol 259: **Nutrition, Growth, and Cancer,** George P. Tryfiates, Kedar N. Prasad, *Editors*

Vol 260: **EORTC Genitourinary Group Monograph 4: Management of Advanced Cancer of Prostate and Bladder,** Philip H. Smith, Michele Pavone-Macaluso, *Editors*

Vol 261: **Nicotine Replacement: A Critical Evaluation,** Ovide F. Pomerleau, Cynthia S. Pomerleau, *Editors*

Vol 262: **Hormones, Cell Biology, and Cancer: Perspectives and Potentials,** W. David Hankins, David Puett, *Editors*

Vol 263: **Mechanisms in Asthma: Pharmacology, Physiology, and Management,** Carol L. Armour, Judith L. Black, *Editors*

Vol 264: **Perspectives in Shock Research,** Robert F. Bond, *Editor*

Vol 265: **Pathogenesis and New Approaches to the Study of Noninsulin-Dependent Diabetes Mellitus,** Albert Y. Chang, Arthur R. Diani, *Editors*

Vol 266: **Growth Factors and Other Aspects of Wound Healing: Biological and Clinical Implications,** Adrian Barbul, Eli Pines, Michael Caldwell, Thomas K. Hunt, *Editors*

Vol 267: **Meiotic Inhibition: Molecular Control of Meiosis,** Florence P. Haseltine, Neal L. First, *Editors*

Vol 268: **The Na^+,K^+-Pump,** Jens C. Skou, Jens G. Nørby, Arvid B. Maunsbach, Mikael Esmann, *Editors.* Published in two volumes: Part A: *Molecular Aspects.* Part B: *Cellular Aspects.*

Vol 269: **EORTC Genitourinary Group Monograph 5: Progress and Controversies in Oncological Urology II,** Fritz H. Schröder, Jan G.M. Klijn, Karl H. Kurth, Herbert M. Pinedo, Ted A.W. Splinter, Herman J. de Voogt, *Editors*

Vol 270: **Cell-Free Analysis of Membrane Traffic,** D. James Morré, Kathryn E. Howell, Geoffrey M.W. Cook, W. Howard Evans, *Editors*

Vol 271: **Advances in Neuroblastoma Research 2,** Audrey E. Evans, Giulio J. D'Angio, Alfred G. Knudson, Robert C. Seeger, *Editors*

Vol 272: **Bacterial Endotoxins: Pathophysiological Effects, Clinical Significance, and Pharmacological Control,** Jack Levin, Harry R. Büller, Jan W. ten Cate, Sander J.H. van Deventer, Augueste Sturk, *Editors*

Vol 273: **The Ion Pumps: Structure, Function, and Regulation,** Wilfred D. Stein, *Editor*

Vol 274: **Oxidases and Related Redox Systems,** Tsoo E. King, Howard S. Mason, Martin Morrison, *Editors*

Vol 275: **Electrophysiology of the Sinoatrial and Atrioventricular Nodes,** Todor N. Mazgalev, Leonard S. Dreifus, Eric L. Michelson, *Editors*

Vol 276: **Prediction of Response to Cancer Therapy,** Thomas C. Hall, *Editor*

Vol 277: **Advances in Urologic Oncology,** Nasser Javadpour, Gerald P. Murphy, *Editors*

Vol 278: **Advances in Cancer Control: Cancer Control Research and the Emergence of the Oncology Product Line,** Paul F. Engstrom, Paul N. Anderson, Lee E. Mortenson, *Editors*

Vol 279: **Basic and Clinical Perspectives of Colorectal Polyps and Cancer,** Glenn Steele, Jr., Randall W. Burt, Sidney J. Winawer, James P. Karr, *Editors*

Vol 280: **Plant Flavonoids in Biology and Medicine II: Biochemical, Cellular, and Medicinal Properties,** Vivian Cody, Elliott Middleton, Jr., Jeffrey B. Harborne, Alain Beretz, *Editors*

Vol 281: **Transplacental Effects on Fetal Health,** Dante G. Scarpelli, George Migaki, *Editors*

Vol 282: **Biological Membranes: Aberrations in Membrane Structure and Function,** Manfred L. Karnovsky, Alexander Leaf, Liana C. Bolis, *Editors*

Vol 283: **Platelet Membrane Receptors: Molecular Biology, Immunology, Biochemistry, and Pathology,** G.A. Jamieson, *Editor*

Vol 284: **Cellular Factors in Development and Differentiation: Embryos, Teratocarcinomas, and Differentiated Tissues,** Stephen E. Harris, Per-Erik Mansson, *Editors*

Vol 285: **Non-Radiometric Assays: Technology and Application in Polypeptide and Steroid Hormone Detection,** Barry D. Albertson, Florence P. Haseltine, *Editors*

Vol 286: **Molecular and Cellular Mechanisms of Septic Shock,** Bryan L. Roth, Thor B. Nielsen, Adam E. McKee, *Editors*

Vol 287: **Dietary Restriction and Aging,** David L. Snyder, *Editor*

Please contact the publisher for information about previous titles in this series.

Dietary Restriction and Aging

DIETARY RESTRICTION AND AGING

Proceedings of the Symposium on the Effects of Dietary Restriction on Aging and Disease in Germfree and Conventional Lobund-Wistar Rats, held in Notre Dame, Indiana, March 27–29, 1988

Editor

David L. Snyder
Lobund Laboratory
University of Notre Dame
Notre Dame, Indiana

ALAN R. LISS, INC. • NEW YORK

Address all Inquiries to the Publisher
Alan R. Liss, Inc., 41 East 11th Street, New York, NY 10003

Copyright © 1989 Alan R. Liss, Inc.

Printed in the United States of America

Under the conditions stated below the owner of copyright for this book hereby grants permission to users to make photocopy reproductions of any part or all of its contents for personal or internal organizational use, or for personal or internal use of specific clients. This consent is given on the condition that the copier pay the stated per-copy fee through the Copyright Clearance Center, Incorporated, 27 Congress Street, Salem, MA 01970, as listed in the most current issue of "Permissions to Photocopy" (Publisher's Fee List, distributed by CCC, Inc.), for copying beyond that permitted by sections 107 or 108 of the US Copyright Law. This consent does not extend to other kinds of copying, such as copying for general distribution, for advertising or promotional purposes, for creating new collective works, or for resale.

Library of Congress Cataloging-in-Publication Data

Symposium on the Effects of Dietary Restriction on Aging and Disease in Germfree and Conventional Lobund-Wistar Rats (1988 : Notre Dame, Ind.)
Dietary restriction and aging.

(Progress in clinical and biological research ; v. 287)
Includes index.
1. Rats—Aging—Congresses. 2. Rats—Diseases—Congresses. 3. Rats—Feeding and feeds—Congresses. 4. Mammals—Aging—Congresses. 5. Mammals—Diseases—Congresses. 6. Mammals—Feeding and feeds—Congresses. I. Snyder, David L. (David Leroy) II. Title.
QL737.R666S96   1988   599.32'33   88-26604
ISBN 0-8451-5137-1

Contents

Contributors xi

Preface
David L. Snyder xv

INTRODUCTION TO AGING AND DIETARY RESTRICTION

1. **Conference Introduction**
   Edward A. Malloy 3
2. **Aging: Where We Have Been and Where We Are Going**
   Jacob A. Brody 7
3. **Overview of the Effects of Food Restriction**
   Edward J. Masoro 27

LIFE SPAN AND PATHOLOGY

4. **The Design of the Lobund Aging Study and the Growth and Survival of the Lobund-Wistar Rat**
   David L. Snyder and Bernard S. Wostmann 39
5. **Spontaneous Diseases in Aging Lobund-Wistar Rats**
   Morris Pollard and Phyllis H. Luckert 51
6. **Effects of Dietary Restriction on Body Composition and Body Size in Germfree and Conventional Lobund-Wistar Rats**
   G.A. Boissonneault, T. Giles, and P.B. Meyers 61
7. **Cardiac Fibrosis in the Aged Germfree and Conventional Lobund-Wistar Rat**
   Gibbons G. Cornwell III and Beverly P. Thomas 69
8. **Effects of Aging, Diet Restriction, and Microflora on Oral Health in Humans and Animals**
   Sam Rosen, Mike Strayer, William Glocker, James Marquard, and F. Mike Beck 75
9. **Age-Related Changes in Organic and Inorganic Bone Matrix Constituents: The Effect of Environment and Diet**
   Satoru K. Nishimoto and Steven M. Padilla 87

IMMUNOLOGY

10. How Does Dietary Restriction Retard Diseases and Aging?
    Richard Weindruch 97
11. Immune Function in Aging Rats: Effects of Germfree Status and Caloric Restriction
    Kara W. Eberly and E. Bruckner-Kardoss 105
12. Effect of Dietary Restriction and Aging on Lymphocyte Subsets in Germfree and Conventional Lobund-Wistar Rats
    Yoon Berm Kim and Alice Gilman-Sachs 117
13. C-Reactive Protein in Aging Lobund-Wistar Rats
    Joan N. Siegel and Henry Gewurz 127

ENDOCRINOLOGY

14. The Effect of Dietary Restriction on Serum Hormone and Blood Chemistry Changes in Aging Lobund-Wistar Rats
    David L. Snyder and Beth Towne 135
15. Age-Related Changes in Adrenal Catecholamine Levels and Medullary Structure in Male Lobund-Wistar Rats
    Nancy P. Nekvasil and Toni R. Kingsley 147
16. Modest Dietary Restriction and Serum Somatomedin-C/Insulin Like Growth Factor-I in Young, Mature, and Old Rats
    T. Elaine Prewitt and A. Joseph D'Ercole 157
17. Changes in Pancreatic Hormones During Aging
    Richard C. Adelman 163

NEUROENDOCRINOLOGY

18. Evidence That Underfeeding Acts Via the Neuroendocrine System to Influence Aging Processes
    Joseph Meites 169
19. Adenohypophysial Changes in Conventional, Germfree, and Food-Restricted Aging Lobund-Wistar Rats: A Histologic, Immunocytochemical and Electron Microscopic Study
    Kalman Kovacs, Nancy Ryan, Toshiaki Sano, Lucia Stefaneanu, Gezina Ilse, and Sylvia L. Asa 181
20. The Morphology of Adenohypophysial Cells in Aging Lobund-Wistar Rats in Tissue Culture: An Ultrastructural Study
    Sylvia L. Asa, Kalman Kovacs, Blair M. Gerrie, Robin E. Baird, and Gezina Ilse 191
21. Dietary Restriction and Hypothalamic Metabolism: A Model of Aging and Food Intake Control
    Roy J. Martin 201

GASTROINTESTINAL PHYSIOLOGY

22. Effect of Dietary Restriction on Gastrointestinal Cell Growth
    Peter R. Holt 211
23. Fatty Acid Compositional Changes in Germfree and Conventional Young and Old Rats
    G. Bruckner and K. Gannoe-Hale 221

NUTRITIONAL BIOCHEMISTRY

24. Functional and Biochemical Parameters in Aging Lobund-Wistar Rats
    Bernard S. Wostmann, David L. Snyder, Margaret H. Johnson, and Shi Shun-di 229
25. Blood Glutathione: A Biochemical Index of Life Span Enhancement in the Diet Restricted Lobund-Wistar Rat
    Calvin A. Lang, Wenkai Wu, Theresa Chen, and Betty Jane Mills 241
26. Cellular Antioxidant Defense System
    Linda H. Chen and Stephen R. Lowry 247

CELLULAR BIOCHEMISTRY

27. An Overview of Age-Related Changes in Proteins
    Morton Rothstein 259
28. The Role of Oxidative Modification in Cellular Protein Turnover and Aging
    Pamela E. Starke-Reed 269
29. Age-Related Molecular Changes in Skeletal Muscle
    Ari Gafni and Khe-Ching M. Yuh 277
30. Dietary Restriction Postpones the Age-Dependent Compromise of Male Rat Liver Microsomal Monooxygenases
    Douglas L. Schmucker and Rose K. Wang 283
31. Stimulation of DNA Chain Initiation by a Protein Factor (NPF-1) From Rat Liver of Different Ages
    Subhash Basu, Adrian Torres Rosado, Shigeo Takada, Satyajit Ray, Kamal Das, Isao Suzuki, and Annie Pierre Seve 289
32. Effect of Aging and Dietary Restriction on the Expression of α_{2u}-Globulin in Two Strains of Rats
    Bo Wu, Craig C. Conrad, and Arlan Richardson 301

Index 311

Contributors

Richard C. Adelman, Institute of Gerontology and Department of Biological Chemistry, The University of Michigan, Ann Arbor, MI 48109 **[163]**

Sylvia L. Asa, Department of Pathology, St. Michael's Hospital, University of Toronto, Toronto, Ontario, Canada M5B 1W8 **[181,191]**

Robin E. Baird, Department of Pathology, St. Michael's Hospital, University of Toronto, Toronto, Ontario, Canada M5B 1W8 **[191]**

Subhash Basu, Department of Chemistry, Biochemistry, Biophysics, and Molecular Biology Program, University of Notre Dame, Notre Dame, IN 46556 **[289]**

F. Mike Beck, Department of Diagnostic Services, College of Dentistry, The Ohio State University, Columbus, OH 43210 **[75]**

Gilbert A. Boissonneault, Department of Clinical Nutrition, University of Kentucky, Lexington, KY 40536 **[61]**

Jacob A. Brody, School of Public Health, University of Illinois at Chicago, Chicago, IL 60680 **[7]**

Geza Bruckner, Department of Clinical Nutrition, University of Kentucky, Lexington, KY 40536 **[221]**

Edith Bruckner-Kardoss, Lobund Laboratory, University of Notre Dame, Notre Dame, IN 46556 **[105]**

Linda H. Chen, Department of Nutrition and Food Science, University of Kentucky, Lexington, KY 40506 **[247]**

Theresa Chen, Department of Pharmacology and Toxicology, University of Louisville School of Medicine, Louisville, KY 40292 **[241]**

Craig C. Conrad, Department of Chemistry, Illinois State University, Normal, IL 61761 **[301]**

Gibbons G. Cornwell III, Department of Medicine, Dartmouth Medical School, Hanover, NH 03756 **[69]**

Kamal Das, Department of Chemistry, Biochemistry, Biophysics and Molecular Biology Program, University of Notre Dame, Notre Dame, IN 46556 **[289]**

A. Joseph D'Ercole, Department of Pediatrics, School of Medicine, University of North Carolina, Chapel Hill, NC 27514 **[157]**

Kara W. Eberly, Department of Biology, Saint Mary's College, Notre Dame, IN 46556 **[105]**

The numbers in brackets are the opening page numbers of the contributors' articles.

xii / Contributors

Ari Gafni, Institute of Gerontology and Department of Biological Chemistry, University of Michigan, Ann Arbor, MI 48109 **[277]**

K. Gannoe-Hale, Department of Clinical Nutrition, University of Kentucky, Lexington, KY 40536 **[221]**

Blair M. Gerrie, Department of Pathology, St. Michael's Hospital, University of Toronto, Toronto, Ontario, Canada M5B 1W8 **[191]**

Henry Gewurz, Department of Immunology/Microbiology, Rush-Presbyterian-St. Luke's Medical Center, Chicago, IL 60612 **[127]**

Timothy Giles, Department of Clinical Nutrition, University of Kentucky, Lexington, KY 40536 **[61]**

Alice Gilman-Sachs, Department of Microbiology and Immunology, University of Health Sciences/The Chicago Medical School, North Chicago, IL 60064 **[117]**

William Glocker, Department of Oral Biology, College of Dentistry, The Ohio State University, Columbus, OH 43210 **[75]**

Peter R. Holt, Division of Gastroenterology, Department of Medicine, St. Luke's Hospital Center, and College of Physicians & Surgeons of Columbia University, New York, NY 10025 **[211]**

Gezina Ilse, Department of Pathology, St. Michael's Hospital, University of Toronto, Toronto, Ontario, Canada M5B 1W8 **[181,191]**

Margaret H. Johnson, Lobund Laboratory, University of Notre Dame, Notre Dame, IN 46556 **[229]**

Yoon Berm Kim, Department of Microbiology and Immunology, University of Health Sciences/The Chicago Medical School, North Chicago, IL 60064 **[117]**

Toni R. Kingsley, South Bend Center for Medical Education, University of Notre Dame, Notre Dame, IN 46556 **[147]**

Kalman Kovacs, Department of Pathology, St. Michael's Hospital, University of Toronto, Toronto, Ontario, Canada M5B 1W8 **[181,191]**

Calvin A. Lang, Department of Biochemistry, University of Louisville School of Medicine, Louisville, KY 40292 **[241]**

Stephen R. Lowry, Agricultural Experiment Station, University of Kentucky, Lexington, KY 40506 **[247]**

Phyllis H. Luckert, Lobund Laboratory, University of Notre Dame, Notre Dame, IN 46556 **[51]**

Edward A. Malloy, University of Notre Dame, Notre Dame, IN 46556 **[3]**

James Marquard, Department of Diagnostic Services, College of Dentistry, The Ohio State University, Columbus, OH 43210 **[75]**

Roy J. Martin, Department of Foods and Nutrition, University of Georgia, Athens, GA 30602 **[201]**

Edward J. Masoro, Department of Physiology, University of Texas Health Science Center, San Antonio, TX 78284 **[27]**

Joseph Meites, Department of Physiology, Michigan State University, East Lansing, MI 48824 **[169]**

Paul B. Meyers, Department of Clinical Nutrition, University of Kentucky, Lexington, KY 40536 **[61]**

Contributors / xiii

Betty Jane Mills, Department of Biochemistry, University of Louisville School of Medicine, Louisville, KY 40292 **[241]**

Nancy P. Nekvasil, Biology Department, St. Mary's College, Notre Dame, IN 46556 **[147]**

Satoru K. Nishimoto, Departments of Orthopaedics, Biochemistry, Medicine, and Laboratory of Connective Tissue, Orthopaedic Hospital, University of Southern California, Los Angeles, CA 90007; present address: Department of Biochemistry, University of Tennessee, Memphis, TN 38163 **[87]**

Steven M. Padilla, Department of Orthopaedics and Laboratory of Connective Tissue, Orthopaedic Hospital, University of Southern California, Los Angeles, CA 90007 **[87]**

Morris Pollard, Lobund Laboratory, University of Notre Dame, Notre Dame, IN 46556 **[51]**

T. Elaine Prewitt, Department of Nutrition and Medical Dietetics, College of Associated Health Professionals, University of Illinois at Chicago, Chicago, IL 60680 **[157]**

Satyajit Ray, Department of Chemistry, Biochemistry, Biophysics and Molecular Biology Program, University of Notre Dame, Notre Dame, IN 46556 **[289]**

Arlan Richardson, Department of Chemistry, Illinois State University, Normal, IL 61761 **[301]**

Sam Rosen, Department of Oral Biology, College of Dentistry, The Ohio State University, Columbus, OH 43210 **[75]**

Morton Rothstein, Department of Biological Sciences, State University of New York at Buffalo, Amherst, NY 14260 **[259]**

Nancy Ryan, Department of Pathology, St. Michael's Hospital, University of Toronto, Toronto, Ontario, Canada M5B 1W8 **[181]**

Toshiaki Sano, Department of Pathology, St. Michael's Hospital, University of Toronto, Toronto, Ontario, Canada M5B 1W8 **[181]**

Douglas L. Schmucker, Cell Biology and Aging Section, Veterans Administration Medical Center, the Department of Anatomy, and the Liver Center, University of California, San Francisco, CA 94143 **[283]**

Annie Pierre Seve, Glycoconjogates, CBM-CNRS, Orleans, France **[289]**

Shi Shun-di, Lobund Laboratory, University of Notre Dame, Notre Dame, IN 46556 **[229]**

Joan N. Siegel, Department of Immunology/Microbiology, Rush-Presbyterian-St. Luke's Medical Center, Chicago, IL 60612 **[127]**

David L. Snyder, Lobund Laboratory, University of Notre Dame, Notre Dame, IN 46556 **[39,135,229]**

Pamela E. Starke-Reed, Laboratory of Biochemistry, National Heart, Lung, and Blood Institute, NIH, Bethesda, MD 20892 **[269]**

Lucia Stefaneanu, Department of Pathology, St. Michael's Hospital, University of Toronto, Toronto, Ontario, Canada M5B 1W8 **[181]**

Mike Strayer, Department of Community Dentistry, College of Dentistry, The Ohio State University, Columbus, OH 43210 **[75]**

Isao Suzuki, Department of Chemistry, Biochemistry, Biophysics and Molecular Biology Program, University of Notre Dame, Notre Dame, IN 46556 **[289]**

Shigeo Takada, Department of Biochemistry, School of Medicine, Tokai University, Kanagawa 259-11, Japan **[289]**

Beverly P. Thomas, Department of Medicine, Dartmouth Medical School, Hanover, NH 03756 **[69]**

Adrian Torres Rosado, Department of Chemistry, Biochemistry, Biophysics and Molecular Biology Program, University of Notre Dame, Notre Dame, IN 46556 **[289]**

Beth Towne, Lobund Laboratory, University of Notre Dame, Notre Dame, IN 46556 **[135]**

Rose K. Wang, Cell Biology & Aging Section, Veterans Administration Medical Center, the Department of Anatomy, and the Liver Center, University of California, San Francisco, CA 94143 **[283]**

Richard Weindruch, Biomedical Research and Clinical Medicine Program, National Institute on Aging, NIH, Bethesda, MD 20892 **[97]**

Bernard S. Wostmann, Lobund Laboratory, University of Notre Dame, Notre Dame, IN 46556 **[39,229]**

Bo Wu, Department of Chemistry, Illinois State University, Normal, IL 61761 **[301]**

Wenkai Wu, Department of Biochemistry, University of Louisville School of Medicine, Louisville, KY 40292 **[241]**

Khe-Ching M. Yuh, Institute of Gerontology, University of Michigan, Ann Arbor, MI 48109 **[277]**

Preface

In 1983 the staff of Lobund Laboratory at the University of Notre Dame began an examination of the effects of dietary restriction on aging in germfree Lobund-Wistar rats. It soon became apparent that a study of this magnitude using germfree animals could never be repeated. Therefore the Lobund Aging Study was expanded to include investigators at other universities in the United States and Canada. The distribution of tissues from young to very old rats was completed in 1987, and in March 1988 more than 30 collaborators gathered at Notre Dame to present their findings. The proceedings of that symposium, published here, mark a halfway point in this unique collaborative effort. Complete analysis and publication of the study results will probably require another year. The ultimate goal will be the integration of the data and ideas of all the collaborators into a universal hypothesis for the action of dietary restriction on aging.

I would like to thank the many outstanding people who contributed to the symposium and these proceedings. The Rev. Edward A. Malloy, President of the University of Notre Dame, opened the symposium with his thoughts on aging and ethics. The keynote speaker was Dr. Jacob A. Brody, Dean of the School of Public Health at the University of Illinois at Chicago. His insights into the problems of human aging now and in the future gave special emphasis to the efforts of the symposium participants who are seeking an understanding of the basic mechanisms of aging. Several notable gerontologists provided critical reviews of research on dietary restriction and the biochemistry of aging. Other collaborators in the Lobund Aging Study, experts in the areas of gnotobiology and experimental gerontology, have provided excellent reviews of specific aspects of aging and dietary restriction in the Lobund-Wistar rat.

Special thanks are due to Dr. Morris Pollard, Director of Lobund Laboratory, and to Dr. Bernard S. Wostmann. Both of these men have given me patient instruction and guidance from the time I started as a neophyte in germfree research to my current efforts as editor of these proceedings. Carolyn Robinson and Arlene Snyder are to be commended for transcribing, typing, and editing several of the manuscripts.

The Lobund Aging Study was supported by the Retirement Research Foundation of Chicago. The symposium was funded by grants from the National Institute on Aging and the Retirement Research Foundation.

I would like to dedicate this book to all the people who have worked at Lobund Laboratory for more than 40 years to develop and maintain the germfree animals that made this study possible.

David L. Snyder

INTRODUCTION TO AGING AND DIETARY RESTRICTION

CONFERENCE INTRODUCTION

Reverend Edward A. Malloy, C.S.C.

President of the University of Notre Dame,
Notre Dame, Indiana 46556

It may seem antithetical to hold a conference on longevity and disease in the spring when all life appears to be newly reborn. But in fact the season lends a certain poignancy to the proceedings. In university settings, especially, with young people everywhere, the season livens our hearts and gives us a sense of the beauty of nature and the wonder of creation. Such a background is ideally (one might almost say poetically) suited to the discussion of the quality of life and the reality of aging. It also concentrates the hope we hold for science and technology to affect the conditions of life in a positive way.

My background is in ethics, and through the years I have had occasion to gather with practitioners of the biomedical and other professions in which technology has played a significant role. In the last few decades we have seen a heightened sensitivity to the way in which technology impinges upon ethics, along with a reappraisal of methodology. It was in the 1960's that Daniel Callahan and some others had the foresight to establish the Hastings Institute in New York -- the first sustained and interdisciplinary endeavor outside of the universities and a few medical centers -- to focus and reflect on the kinds of issues you are considering here today.

From that initiative, of course, many other institutes and centers have developed, and the net result is that we have become much more sensitive to the ethical implications of our behavior.

To live in a pluralistic culture where people of good will can disagree on a fundamental level may seem to play to disorder. But in fact it provides us with an opportunity to grapple with concrete issues in an interdisciplinary way, to determine what our national priorities ought to look like, and to ascertain what the proper mix of the public and private sectors should be in shaping the quality of our lives. Such interdisciplinary efforts are a common thing on a campus such as Notre Dame's today, and I am thankful for that -- thankful to have a combination of people with the expertise and background that you hold in common.

The desire to live a long life is universal. But I remember a little section from Gulliver's Travels in which Gulliver, in the Land of the Houyhnhnms, meets a race of people who live to extreme old age, and after that encounter he comments: "They were the saddest people I ran into in my journey." The saddest because they simply had their life span extended while the quality of that life was not keeping pace.

How long would each of us like to live if we had control over our destiny, which we only do in a relative way? And what would we expect with each day? Would our value system change? Would we assume the tasks that we had put off until we had the leisure to enjoy them? Ultimately, would our lives be better for having been longer?

Daniel Callahan published a book recently which suggested that, increasingly, we are going to face difficult choices about the provision of health care. Is our technological capacity to extend life, through support systems, ultimately counter-productive? At some point, when people reach extreme old age, should we say, "Enough!"? The use of this technology, this capacity, has precipitated a broad-ranging debate not only within the old categories such as euthanasia (whether direct or indirect), but regarding the technology itself, the definitions we apply to life and death, and the individual's right to live and to die with dignity.

So, the span of a life is of no particular worth in gauging its quality. I know people who seem to give up when they are in their 40's and 50's, while others, in their 90's are vital, interesting, insightful, and wise. Too many factors are involved to make such things predictable -- one's genetic makeup, environmental circumstances, emotional and mental health, physical durability, some inner desire and drive. Can we affect these things? And in doing so, can we offer something to the broader world?

It seems to me that one of the purposes that a university such as Notre Dame can serve in its natural and social sciences, in its humanities, its philosophy and theology -- in all of its endeavors -- is to apply value-oriented language to the tasks and priorities that can enhance life in our society. That is, we can work on a practical level toward an ideal perception of our existence on the planet -- not a perfect world, but one in which society can share the fruits of its knowledge in an ethical context, providing the hope and dignity that ought to be the birthright of all humans.

There are no easy or cheap solutions. There are problems of distribution; and all of us are troubled by the high cost of scientific research today. We are struggling to find the right kind of interplay between the private sector and the academy.

But by gathering groups like this together, by focusing your attention on an interrelated set of issues, you can help all of us come to grips with these questions.

In the meantime, while you are here, I hope you are not so busy that you don't have a chance to walk around the campus. We have two beautiful lakes. And if you want an overview, you might go up to the top floor of the library, walk around the periphery, look out the windows. Things are growing green; the flowers will be blossoming soon. Maybe by pulling away from the struggles and mental anguish of trying to come to grips with an issue such as the one you are dealing with, a certain balance may enter the conversations.

It's great to have you here; I hope that you will return to be with us soon. May you have a wonderful conference.

AGING: WHERE WE HAVE BEEN AND WHERE WE ARE GOING

Jacob A. Brody, M.D.

Dean, School of Public Health, University of Illinois at Chicago, Box 6998 (M/C 922), Chicago, Illinois 60680

Dr. Jacob Brody:

It is an honor for me to be here and have this opportunity to address you. When Father Malloy was discussing the Callahan Book and such topics, I thought wouldn't it be nice to just go on with that philosophical line and put down what I had prepared. What I have prepared is in a sense, the ingredients which thrust up the philosophical question - The overall dimensions of growth of elderly population over this century and projected into the next century. It seems that the noble words of the patriot Patrick Henry, "Give me liberty or give me death" was a minority opinion. The majority of people are opting increasingly for longer life, and we have provided the mechanism for its achievement.

The best way to explain or to focus on the enormity of what's happened is by considering that in the year 1900, only 25% of all deaths were among people 65 and over. By 1985, 75 percent of deaths occurred after age 65. Fully 30% occurred after age 80, and 20% occurred after age 85. Between now and 2020, probably 50% of deaths will be occurring in those over 80. This means we are going to be old. The research being conducted by this group may improve those years and it is important because we are going to make it into those years.

We all are aware that the greatest health and social concerns are congregated around the last year or two of life. We are now talking of a society in which the last year or two of life is going to be well into the 70's. To have this shift occur in so short a time has presented us with a set of unknown and never-experienced circumstances. Among those 65 and over, half of the people have some form of arthritis, a third have some sort of hearing deficit, 40% have elevated blood pressure, and 15% severe visual impairment. By age 80, 20% of our population have Alzheimer's Disease and 20% are in nursing homes. This describes the current situation. But what of the future. We're not going to be curtailing life - quite the contrary. We're extending it. Our challenge will be to provide the strata on which we can go into an old age without these debilitating conditions, most of which are not killers, but vastly detract from the quality and dignity of aging. We don't have a quick fix. We are going to react, and react and react and pragmatically learn the best ways through research and social strategies to make an aging world a better place.

The first graph (Figure 1) shows the decline in mortality since 1900 and as you can see, it's been remarkable. An instructive insight is that half the gain in this century was accomplished by 1920. It wasn't medical or social manipulations that did it. It was probably more affected by the standard of living, sanitation, urbanization, and nutrition. Not seen on the curve is what was going on with the population 65 and over. In those first 20 years of this century, there was virtually no decline in the death rate among those 65 and over. Remarkably, however, by 1950, the half way point for reduced mortality for those 65 and over was reached. Now what was going on between 1920 and 1950 to suddenly lower the death rates among those 65 and over? Social Security was introduced during those years. Also occurring were Prohibition, the Great Depression, World War II, and the Yankees won most of the World Series. We really don't know what caused the remarkable decline in deaths that again preceded the introduction of any very important applications of medical or social technologies. The next surprise was that progress apparently ceased and from 1950 to 1968, the death rates didn't change. That's very distressing because once you get a trend going, you don't like to see it go away. But it stopped, and then lo-and-behold, it started again more rapidly than ever, and is persisting. Mortality from heart disease and stroke have declined since 1968, but we really don't know why many of

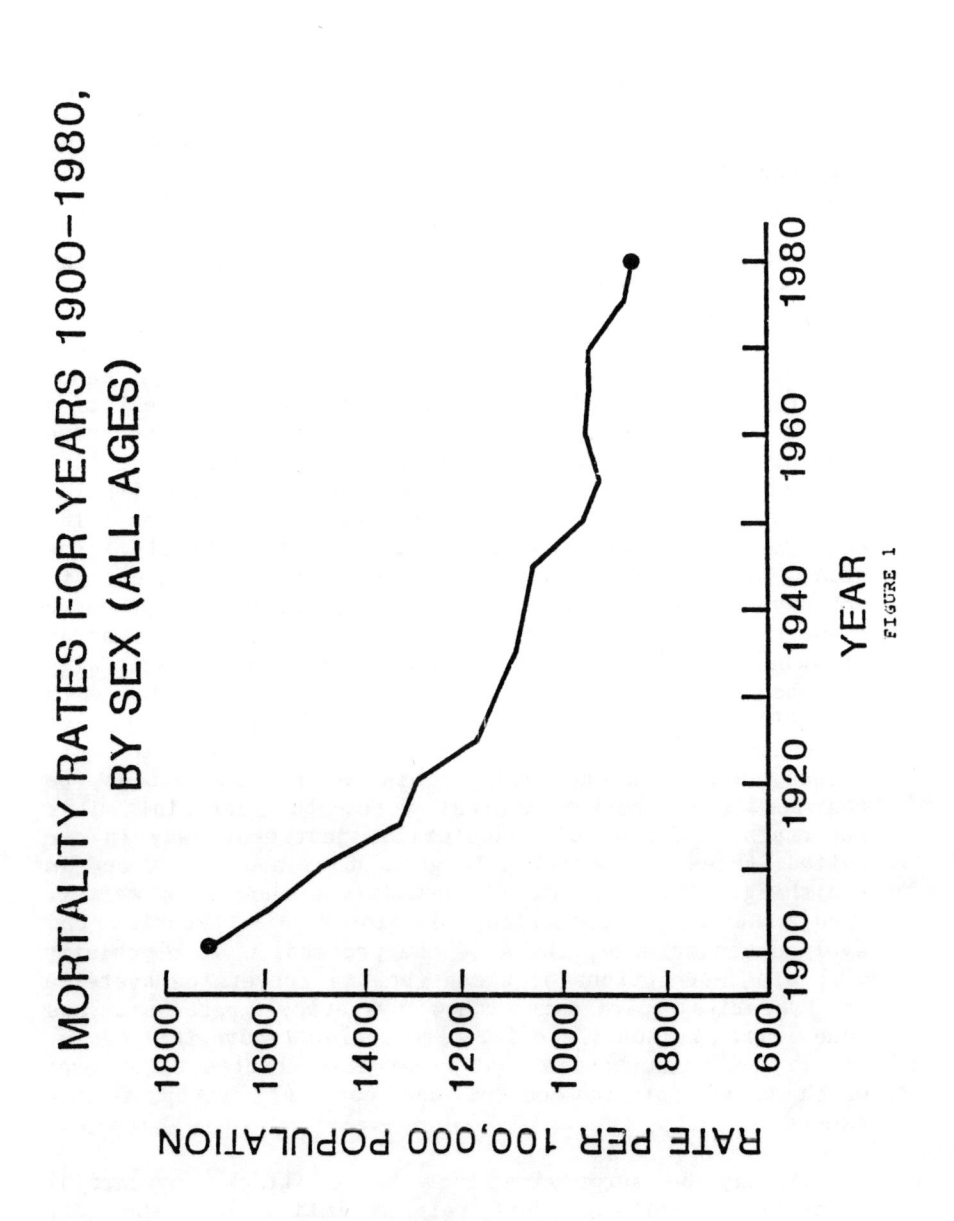

MORTALITY RATES FOR YEARS 1900–1980, BY SEX (ALL AGES)

FIGURE 1

these changes occurred. Patterns of health and disease are apparently somewhat whimsical.

To illustrate how difficult it is to understand why, for instance, heart disease suddenly started to decline in 1968, let me give an example. Figure 2 is for tuberculosis from the year 1860 to 1950. Tuberculosis was the number 1 cause of death in the United States and Northern Europe in 1900, but it had been declining since 1860. We didn't even know that it was caused by a bacillus until 1880; the tuberculin test came in around 1900. The first important therapeutic intervention occurred in the 40's with isoniazide chemotherapy. There's an important thing to be learned here. Disease conditions can change for reasons that are certainly not apparent. There are numerous important examples of this. Measles was a major killer until about 1910 or 1920. There was a manifestation called the black measles that probably very few of you have seen or heard of. Patients would die with hemorrhage into the skin before the rash. Scarlet fever was the leading cause of death from streptococcus until about 1905 or 1910, and then, for some reason, it just modified itself. The strain of streptococcus that became the most prevalent did not have the ability to produce scarlet fever. It produced rheumatic fever, streptococcal sore throat, and an array of other things, but one entire disease was essentially left behind. And these great shifts don't only occur with infections. For some reason since 1950, death rates for accidents - vehicular and non-vehicular, and for all causes, have been declining. This may to some extent, be the result of better medical care, but certainly it's surprising. Cancer of the stomach, just went away in the United States. It started to go down by about 1920 and is vanishing. Even psychiatric conditions show this whimsy. Freud made his reputation, developed his theories and evolved the science, the art, the profession of psychiatry with his descriptions of women who had conversion hysteria or hysterical paralysis. Now hysterical paralysis has gone away. We don't see it. Freud didn't cure it. Again, I'm trying to emphasize that remarkable changes occur over which we're not in control and can only guess at the causes.

It may be surprising that heart attack, myocardial infarction, crushing chest pain radiating down the left arm, which many of us have had or seen, at least on TV,

Figure 2
RESPIRATORY TUBERCULOSIS IN THE TOTAL POPULATION

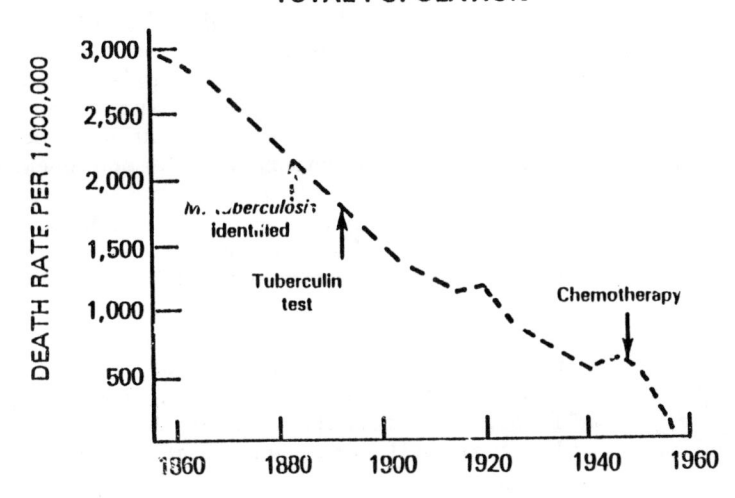

[Graph: Death Rate per 1,000,000 vs. Year (1860–1960), showing declining dashed curve from ~3,000 to near 0. Annotations: "M. tuberculosis identified", "Tuberculin test", "Chemotherapy".]

Source: Mean annual death rates from various diseases, England and Wales, 1860-1960. From Winikoff B: *Nutrition and National Policy*, Cambridge, MIT Press, 1978, pp. 444-445.

was not described in the medical literature until 1912. It may have been missed by doctors, or perhaps they just were calling it indigestion. Everyone likes to bash doctors a little. But some of the greatest clinical sources like Shakespeare, Dosteovsky, Dickens and the Bible do not describe this very dramatic, very easy to describe condition. It probably was occurring, but at a very low rate that didn't attract much attention. Then why the sudden change? It had to occur because something provided it with a fertile ground, a chance for it to flourish. An important cluster of events occurred at the turn of the century relating to what we all now know as risk factors for heart disease. We increased our consumption of dietary fat; the choice of smoking products became the cigarette and the nature of the labor force changed remarkably causing men to perform much less physical work. These three things and many others, conspired to produce this incredible surge of heart attacks. Diseases of the heart account for one half of all deaths and earlier this century, we introduced a major killer which accounts for anywhere between 20 and 50% of those heart disease deaths.

I suspect that the line in Figure 1 would really be a straight line, but the huge surge and subsequent decline in heart disease altered the curve. The flattening out was the intercept of the heart disease epidemic interfering with life extension that the various social and medical and unknown factors were providing for us. It is important to recognize that something can appear suddenly and be of such magnitude, as to alter the patterns upon which both scientists and politicians rely for understanding and decision making. Most of the health scientists, demographers and politicians and policy makers, assumed that between 1950 and 1968, humans had reached their maximum life expectancy of about 68 years and Social Security policies and Medicare was created under that assumption. Had we been wiser, we would have realized that myocardial infarct was an opportunistic event in this century. As soon as we modified risk factors such as smoking, exercise, diet, to a relatively small degree, mortality from heart disease slipped away, and thrust additional years onto life.

Life expectancy has been increasing for a long time, and probably is related to improved nutrition and decline in infectious diseases particularly during the developmental years of childhood. Analogous is the phenomenon that we are growing taller. It has been documented going back before 1850, that each generation is increasing in height. And we don't know where it is going to stop. Probably we are more closely approaching the genetic potential for height. This must have a lot to do with early nutrition and the avoidance of frequent infection, during a phase in which growth cells are multiplying. Is increasing height related to longevity? I would say that there is a strong probability that either overlapping or identical factors are propelling us into longer and taller lives. In the United States, as I mentioned, we put great weight on the fact that a large segment of our recent gains in life expectancy has resulted from the decline of cardiovascular disease. However, in two countries in which people are living longer than we are, (Japan and Sweden), they haven't given up smoking, they don't jog, they eat fatty diets and the cardiovascular diseases in those countries are not declining. So there are many ways in which we can live longer.

But, of course, the comment has already been mentioned by Father Malloy. What we are really concerned about is not living longer, but leading healthier lives. We definitely have more older people. Are we simply placing a larger number of people, at ages at which they are very high risk for numerous undesirable consequences, or are we prolonging the good parts of life? Mae West's opinion was that "Too much of a good thing is wonderful!" There's a very important idea that was introduced recently called active life expectancy. It is quantifiable. It means the years you live without depending on anyone for help in a defined array of normal activities. The problem facing us since we are definitely increasing active life expectancy, is, are we increasing active life expectancy as much as we're increasing overall life expectancy? If you look at current data, you are drawn into pessimism. For every year, of added active life expectancy, we add about three to four years of compromised health. We've got to work on that. That's the burden - the sense of fear that people have of growing old, and of course, that a society has of having to take care of them. We must find out how to achieve additional years of active life expectancy at a greater rate than total life expectancy, by postponing sickness to a time much closer to death. We all would like to go out like the wonderful one-horse shay, perfectly healthily until a sudden final moment. The real question is, Are we as a society willing to pay for our success? Right now, we're just shocked by the sticker price. It's costing an awful lot with the improvement of technology and increasing longevity. But as a society, we have to accept the decision, we have to pick up the tab. This is becoming urgent because the baby boom turns 65 in the year 2010.

And now a view to the future (Figure 3). This is a classic demographic pyramid, not quite a pyramid, because in the middle we see the baby boom bulge. This is for 1980: on the right - women, on the left - men, by age. In developing countries, we have true pyramids. The next figure (Figure 4) illustrates what's going to happen by the year 2000 - you can see the bulge is higher, and the pyramid is becoming a little squarer because the top is wider. You'll notice on the right that women live longer than men, and the differences increase with age. By the year 2030 (Figure 5), we have practically a squared pyramid. In other words, we're thrusting up a higher and higher percentage of our population into the older age groups and this puts pressure on all systems.

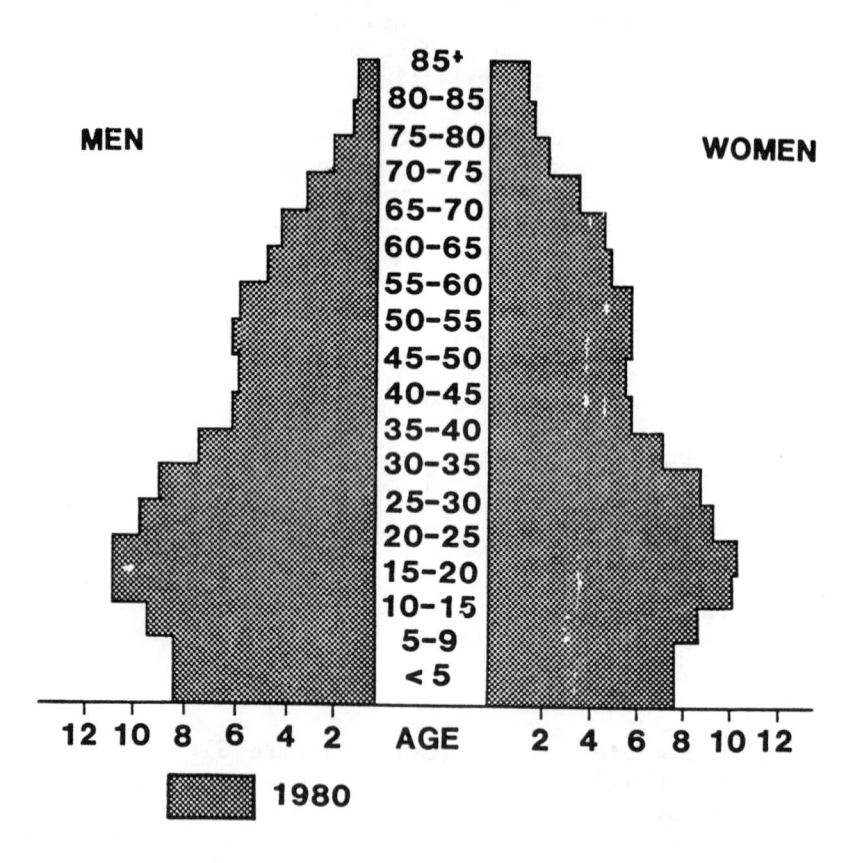

TOTAL POPULATION (IN MILLIONS) BY AGE GROUP AND SEX

FIGURE 3

Aging: Where We Are Going / 15

TOTAL POPULATION (IN MILLIONS) BY AGE GROUP AND SEX

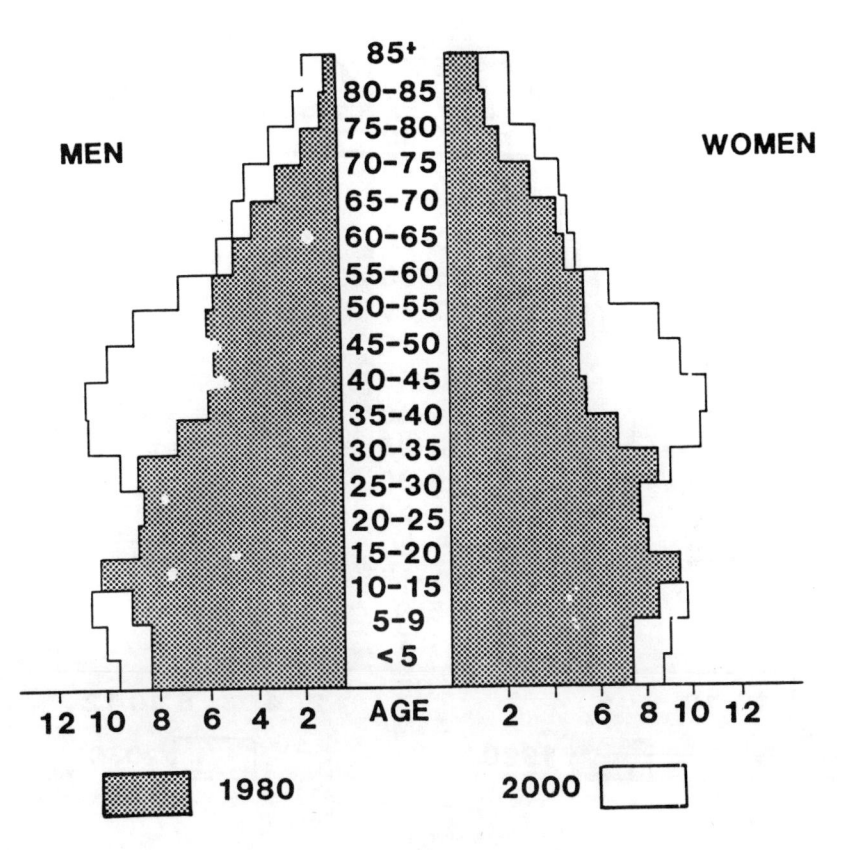

MEN — WOMEN

Age groups: 85+, 80-85, 75-80, 70-75, 65-70, 60-65, 55-60, 50-55, 45-50, 40-45, 35-40, 30-35, 25-30, 20-25, 15-20, 10-15, 5-9, <5

12 10 8 6 4 2 AGE 2 6 8 10 12

▓ 1980   2000 ☐

FIGURE 4

TOTAL POPULATION (IN MILLIONS) BY AGE GROUP AND SEX

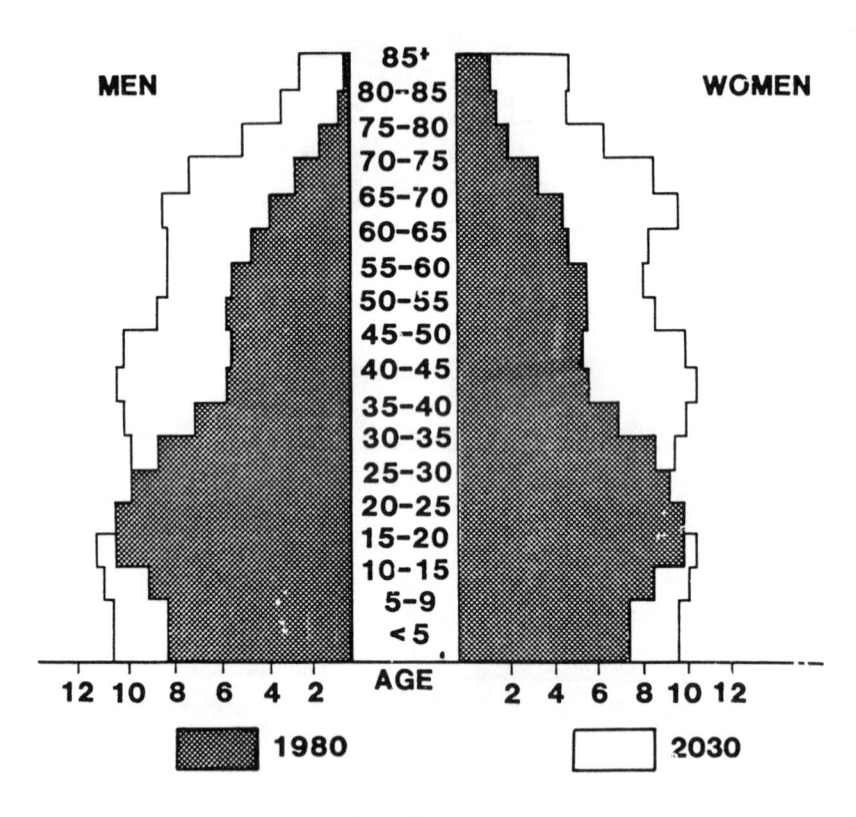

FIGURE 5

I've chosen three specific examples to illustrate the consequences using conservative assumptions. First, we see what's going to happen with hip fractures (Figure 6). In the United States, now there are about 250,000 hip fractures a year. By the year 2000, simply because of the way that population is shifting, and the fact that the median age for hip fracture is 79, we'll be pushing a lot more people through that 79th year. Therefore, by the year 2000, we should have 350,000 hip fractures, and by the year 2020, we'll have a half million hip fractures a year. This next slide is much more discouraging. Figure 7, represents Alzheimer's Disease and other dementias. For hip fracture, we were talking about 250,000 per year in 1985. For the dementias, there are about 2.5 million patients, with median age of about 80. By the year 2000, there should be 3,800,000 patients. By the year 2020, 5 million, and, staggering as it seems, by the year 2050 almost 9 million patients with Alzheimer's. Alzheimer's Disease involves the whole family and social network, not just the affected person. The median age for Alzheimer's is 80, and the average patient survives 7 to 10 years from diagnosis and cost estimates are in the 50 billion dollar range for the country.

This last graph is of nursing home admissions (Figure 8). It must be emphasized that 95% of the population 65 and over do not live in nursing homes, and are doing quite well. Seventy-five to 80% of the people over age 80 do not live in nursing homes, and are doing quite well. Currently there are about 1.5 million people over age 65 in nursing homes, and by the year 2000, they'll be 2.2 million. By the year 2020, there will be 4 million simply projecting the current distribution and use of nursing homes in the U.S. today. In Holland and in Canada, instead of 5%, they have 10% of the elderly in nursing homes. Nursing homes are elastic in the social framework. Because of low birth rates and the fact that women working has increased between 1940 and 1970 from about 10% to about 60%, we have fewer children to take care of the elderly, and the women are not at home. Therefore, there is greater pressure for alternative solutions for taking care of the elderly, including an increased use of nursing homes.

PROJECTED NUMBER OF HIP FRACTURES ANNUALLY IN THE U.S. BY AGE: 1980-2050

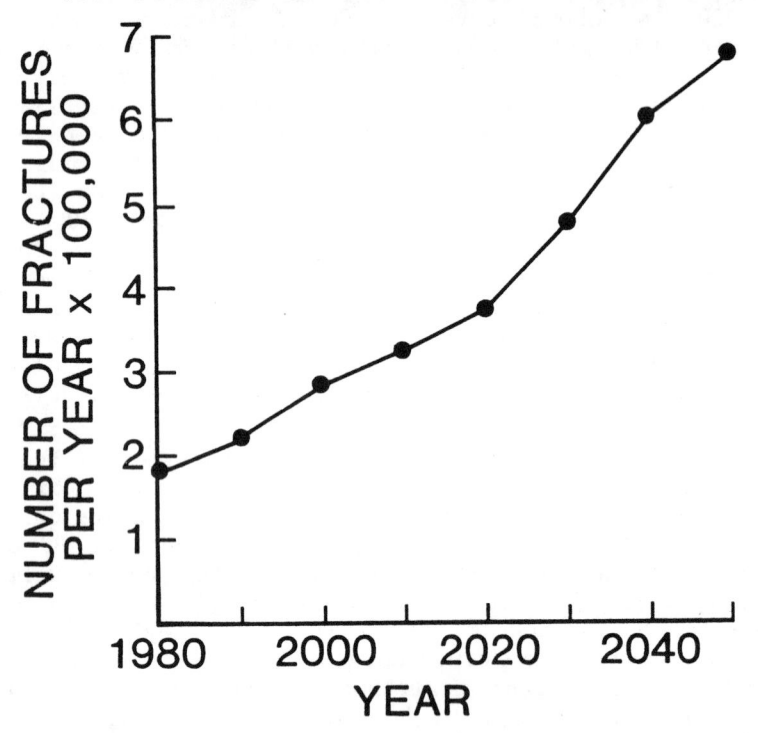

"Source: NCHS and U.S. Bureau of Census projections"

FIGURE 6

PROJECTED NUMBER OF DEMENTED PERSONS IN THE U.S. BY AGE: 1980-2050

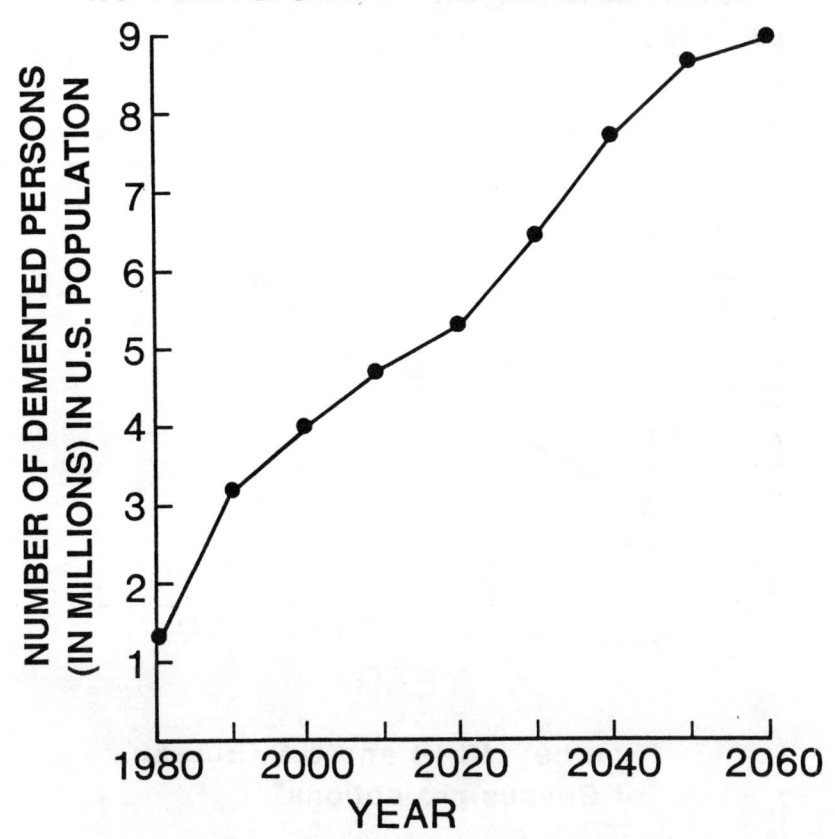

"Source: NIA prevalence estimates and U.S. Bureau of Census projections"

FIGURE 7

PROJECTED NUMBER OF NURSING HOME RESIDENTS IN THE U.S. BY AGE: 1980-2050

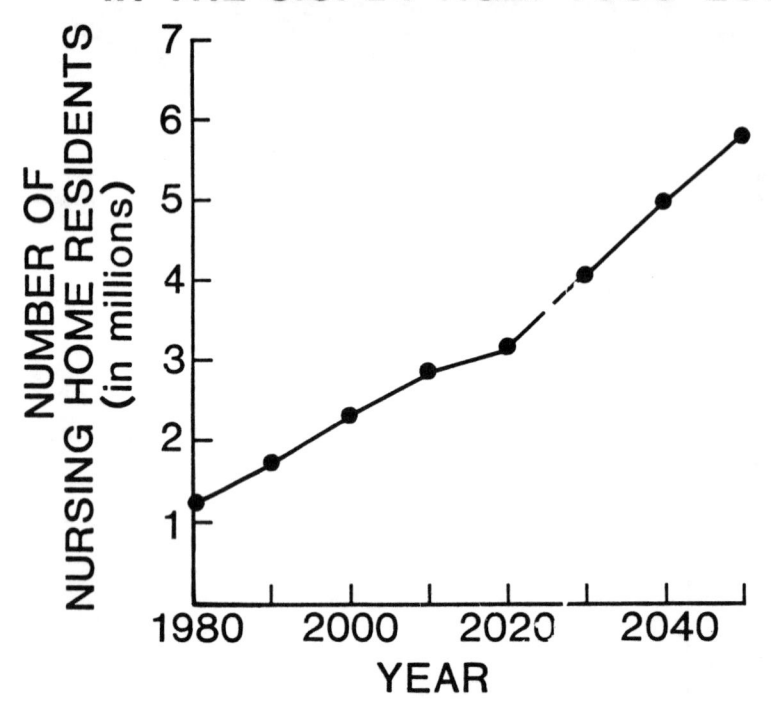

"Source: NCHS and U.S. Bureau of Census projections"

FIGURE 8

Now, does all this gloom that I've been throwing out mean that we're doing something very wrong? I would hasten to say NO! We're doing something very right and we don't know how to handle the implications. Are we ready to change our national policies? And the question is - change to what? As H.L. Menken states in his insightful Metalaw, "For every human problem there is a neat plain solution, and it is always wrong." Compared to the rest of the world, we are not doing badly. We reviewed data from the 17 most-developed countries, reporting age-specific mortality to the World Health Organization since 1950 (Table 1). Figure 9 shows life expectancy at birth from 1950 through 1985. The J is Japan, the U is the United States, and I'll use those two countries as the focus for my point. You can see that the Japanese went from lowest to highest life expectancy at birth during this period. This graph has to be a source of pride to Japanese. The Americans have been in the lower half of this curve over the entire 35-year period. The reasons have to do with ghettos and teenage pregnancies, and a high infant mortality rate. At age 65 (Figure 10), the bottom point again in 1950 was Japan and again, the top point in 1985 was Japan. During this entire period, the United States has been among the leaders in the world. We have been doing better than most countries at age 65. Remarkably at age 75, we're doing better than any other country (Figure 11).

This brings us to the discussion concerning the advisability for the U.S. to adapt a national health policy that will lead us by the hand, out of our aging problems, which I dwelt upon for so long in this talk. I would doubt it. There is much we really don't know, but most other countries have national health plans but none of them are doing better than we are. If we were to accept a radical change, how would we measure our success if we're already living longer? What I'm really suggesting and I should probably restate Menken's law, "For every human problem there is a neat plain solution and it is always wrong." We don't have a quick fix. Pragmatism is what I spoke of earlier and I believe it remains our best ally. We have to introduce things which may be costly, and surely will all have good and bad sides. We hope we will be able to measure the difference. We're very weak at being able to measure the quality of life. We can measure time of death. We'll have to improve our evaluation techniques while trying new things. Some will work, some will only work in certain places.

TABLE 1

Selected Countries, Letter Codes & Population

Letter Code	Country	Population in 1977 (millions)
A	Australia	13.8
B	Belgium	9.8
C	Canada	23.3
D	Denmark	5.1
E	England & Wales	49.3
F	France	53.3
G	Greece	9.2
H	Netherlands	13.8
I	Italy	56.5
J	Japan	113.8
N	Norway	4.0
S	Sweden	8.3
U	USA	215.6
W	West Germany	61.6
Z	Switzerland	6.3

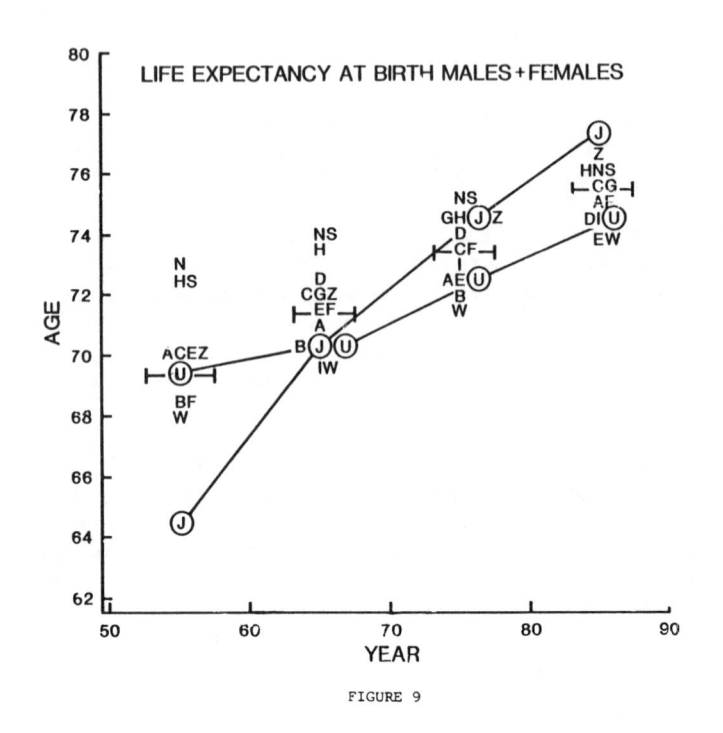

FIGURE 9

Aging: Where We Are Going / 23

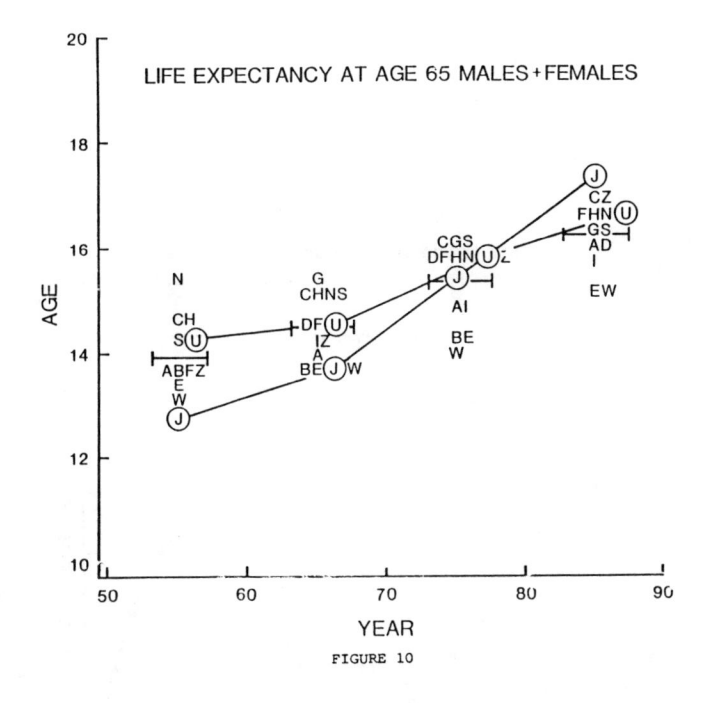

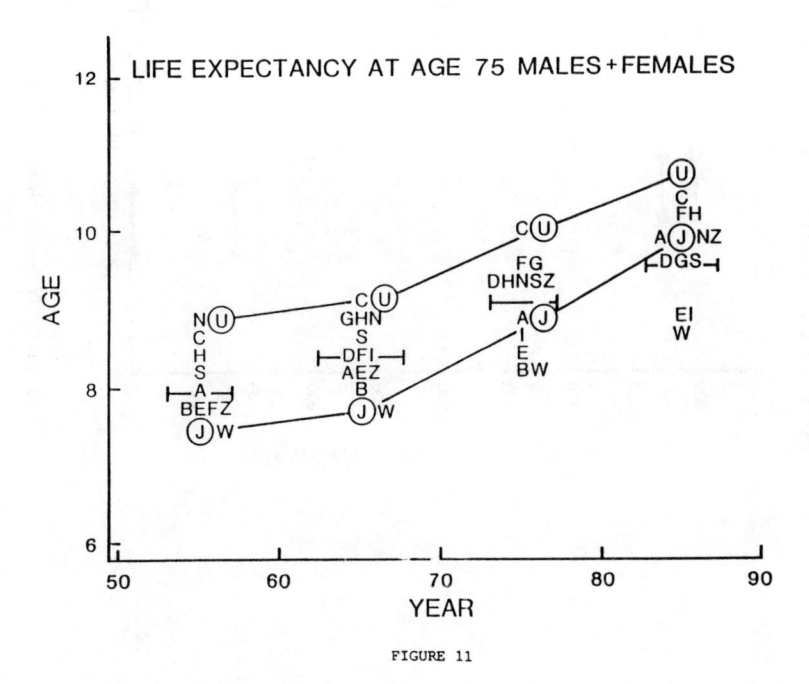

FIGURE 10

FIGURE 11

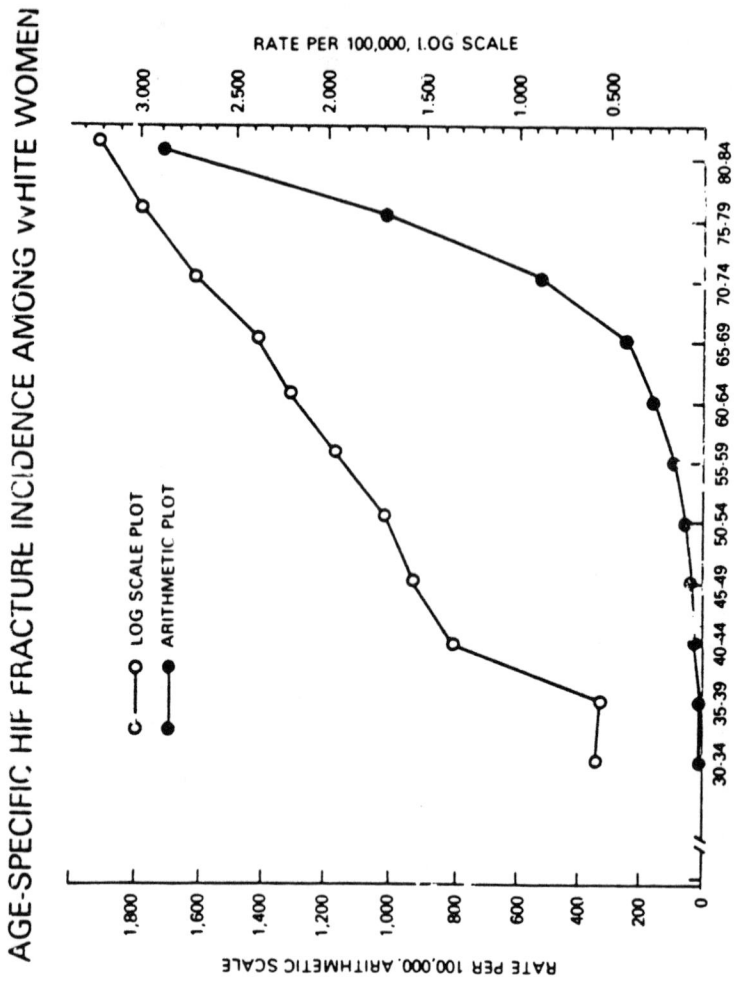

AGE-SPECIFIC HIP FRACTURE INCIDENCE AMONG WHITE WOMEN

○——○ LOG SCALE PLOT
●——● ARITHMETIC PLOT

Source: National Hospital Discharge Survey 1975 to 1979.

FIGURE 12

Our goal is prevention of the downside of aging, and postponing the most negative aspects. I am optimistic that thorough research, we can learn some fundamental things, which will greatly enhance later life. This curve (Figure 12) is of hip fracture in U.S. white females - the lower curve is arithmatic, the top curve is logarithmic, but showing the exact same data. It's just easier to make my point using the upper curve. Notice that the hip fracture rate rises very sharply at about age 40 - a seven fold increase! After that sharp inflection, the rate just continues to rise steadily with age, doubling every 6 years. So on this graph, half the cases will occur beyond the next to last point. With hip fractures, we know they're somehow related to osteoporosis, and osteoporosis is related to bone metabolism. Thus, through better understanding of the osteoblastic mechanisms of the laying down of the bone and of the preservation of the structure to prevent osteoporosis, we could promote the system and postpone the inflection at age 40, until age 46. Then the median case will be at age 84. Since life expectancy for women is about 78, most hip fractures would occur after death, and that's very good prevention. We may actually be able to prevent half of the hip fractures if we could postpone the events leading to osteoporosis by 6 years.

Nutrition, because it enters into so many systems, will be probably the most valuable ally in finding those conditions which are amenable to postponement. The goal is to postpone bad events to occur at ages very late in life, or, hopefully, after death. Thank you. May your conference go well!

OVERVIEW OF THE EFFECTS OF FOOD RESTRICTION

Edward J. Masoro

Department of Physiology, University of Texas Health Science Center, San Antonio, Texas 78284-7756

INTRODUCTION

This overview of the effects of food restriction on the aging of rodents is based on research carried out in my laboratory in which specific pathogen-free male Fischer 344 rats were restricted to 60% of the mean ad libitum food intake. This parochial coverage is permissible because the major effects of food restriction have been shown to occur in a wide variety of strains of mice and rats of both sexes in studies that have involved many different dietary programs and regimens (Holehan and Merry, 1986). What all these different studies have in common is undernutrition without malnutrition (Walford, 1985). The conclusion to be drawn from these diverse studies is that undernutrition without malnutrition in rodents retards the primary aging processes. However, the mechanisms underlying this retardation remain to be determined.

Physiological Processes

There are many age-changes in the physiological processes and systems (Masoro, 1986) and in rodents food restriction delays or prevents most (Masoro, 1985). In the case of male Fischer 344 rats we have observed food restriction to retard the following: the increases in the plasma concentrations of cholesterol, triglyceride, phospholipid, and ketone bodies (Liepa et al, 1980; Masoro et al, 1983); the loss in ability of adipocytes to respond to the lipolytic action of glucagon (Bertrand et al, 1980b) by influencing receptors (Bertrand et al, 1987) and of epinephrine (Yu et al, 1980) by influencing

postreceptor processes (Bertrand et al, 1987); the increase
in the plasma concentration of parathyroid hormone (Kalu et
al, 1984), calcitonin (Kalu et al, 1983) and insulin (Masoro
et al, 1983); the decrease in spontaneous locomotor activity
(Yu et al, 1985); the decrease in tension development by the
smooth muscle of aortic strips (Herlihy and Yu, 1980). Other
investigators have shown many other age-associated
physiological changes to be blunted by food restriction in
wide variety of strains of mice and rats (Holehan and Merry,
1986).

However, food restriction does not retard all age changes
in physiological processes. We found that the age-associated
increase in systolic blood pressure was not influenced by
food restriction (Yu et al, 1985). Moreover, the age-
associated decline in the ability of β receptor stimulation
to cause relaxation of arterial smooth muscle was found to be
enhanced by food restriction (Herlihy and Yu, 1980). There
may be many other examples of the failure of food restriction
to retard age-changes in physiological systems that have not
been reported because of the difficulties encountered in
publishing negative findings. If so, it is unfortunate
because information on what food restriction fails to do may
be important for the pursuit of the mechanism of action of
food restriction and in the quest for knowledge of the
primary aging processes.

Pathologic Lesions

The major age-associated disease process in the male
Fischer 344 rat is chronic nephropathy (Maeda et al, 1985).
The severity of these kidney lesions progressively increases
through most of the life span and renal failure is a major
contributor to the death of more than 50% of the rats fed our
standard semisynthetic diet ad libitum. Food restriction
almost totally prevents this age-associated disease process.
It has been claimed that protein is the major dietary factor
in the development of glomerular sclerosis in humans and rats
(Brenner et al, 1982) and thus it is reasonable to suggest
that it is the restriction of protein by the food restriction
regimen that is responsible for the prevention of chronic
nephropathy. This has not proven to be the case. Although
restriction of protein without restriction of calories (Maeda
et al, 1985) did retard the progression of chronic
nephropathy as did replacing the casein in the semisynthetic
diet with soy protein (Iwasaki et al, 1988a), neither was as

effective as food restriction. Moreover, restriction of calories without restriction of protein also markedly retarded the progression of chronic nephropathy (Masoro et al. 1988). The conclusion to be drawn is that it is probably calorie restriction which is primarily responsible for retarding chronic nephropathy in food restricted male Fischer 344 rats.

Food restriction also almost completely retards the progression of cardiomyopathy, another major age-associated disease process in male Fischer 344 rats (Maeda et al, 1985). Food restriction delays the appearance of neoplastic disease in male Fischer 344 rats but does not reduce its prevalence at the time of spontaneous death (Maeda et al, 1985). It must be pointed out, however, that the food restricted rats were much older at the time of spontaneous death than the ad libitum fed animals. In this strain of rat, it appears that food restriction has a less powerful effect on neoplastic disease than on other age-associated disease processes.

Longevity

In our laboratory, food restriction was found to markedly and reproducibly increase both the life expectancy and life span (Yu et al, 1982; Yu et al, 1985). Moreover, it appears that this increase in longevity probably results from caloric restriction since the equivalent restriction of protein (Yu et al, 1985) or fat (Iwasaki et al, 1988b) or mineral (Iwasaki et al, 1988b) without restriction of calories had either a small or no effect on life expectancy or life span.

Evidence That Food Restriction Retards the Aging Process

Since the nature of the primary aging processes is not known, it is not possible to unequivocally establish that food restriction retards the primary aging processes. There is, however, strong circumstantial evidence for such an action.

The strongest evidence is the fact that it markedly increases life span. Although increases in life expectancy could be due to many factors, an increase in life span almost certainly results from the slowing of aging processes (Sacher, 1977). The great breadth of the effects of food restriction on the age changes in physiological processes also provides strong evidence for an influence on primary aging processes underlying the myriad of secondary and further removed functional changes that characterize aging. Similarly, the fact that food restriciton retards the occurrence and/or progression of most age-associated diseases indicates that it acts by influencing primary aging processes rather than by directly modulating the pathogenesis of specific diseases.

Mechanism of Action of Food Restriction

Research in our laboratory and in many others as well is being focused on the mechanisms by which food restriction retards the aging processes. The reasons for this are two. First is the insight that such knowledge can yield on the nature of the primary aging processes. The second is the data base it may provide in regard to developing interventions in human aging.

Our research using the male Fischer 344 rats has ruled out as important factors, the three major hypotheses proposed for the mechanism by which food restriction retards the aging processes. McCay et al (1935) proposed that food restriction increases life span by retarding growth and development. Our work with the Fischer 344 rat (Yu et al, 1985; Maeda et al, 1985) showed that food restriction started at 6 months of age was as effective as that started at 6 weeks of age in extending life span, delaying age-associated physiological changes, retarding the progression and occurrence of age-associated diseases. Berg and Simms (1960) hypothesized that food restriction retards the aging processes by reducing body fat content. In our studies with the male Fischer 344 rat (Bertrand et al, 1980a), no correlation was found between body fat content and length of life in ad libitum fed animals but in the food restricted rats there was a positive correlation. Thus, it was concluded that although food restricted rats were leaner than ad libitum fed rats, this leanness does not play a causal role in the increase in life span. Sacher (1977) postulated that food restriction retards the aging process by decreasing the metabolic rate. This was embraced by many because of the evidence that reducing food intake reduces the metabolic rate and because of the fact that such a mechanism should reduce the generation of oxygen free radicals (Harman, 1981). Direct measurement of oxygen consumption over twenty-four hour periods in ad libitum and food restricted Fischer 344 rats under usual living conditions revealed that prolonged food restriction does not decrease metabolic rate per unit lean body mass or per unit "metabolic mass" (McCarter et al, 1985). The lean body mass is decreased by food restriction so that neither metabolic rate nor food intake is decreased per unit of "metabolic mass" (Masoro et al, 1982). Thus the hypothesis that food restriction retards the aging processes by decreasing the metabolic rate should be discarded.

Our findings with the male Fischer 344 rats challenge the classic view that food restriction influences the aging processes by reducing the intake of calories or other nutrients per unit of metabolic mass. It appears that it is the nutrient intake per rat rather than per unit metabolic mass that is important. Our hypothesis is that food restriction is coupled to the aging processes by the endocrine and/or nervous systems which influence the biochemical processes underlying aging in a spectrum of target tissues. Our research is now focused on the nature of the coupling system and on the modulation of biochemical processes in the target tissue.

Our guide to potential couplers is the information indicating an involvement of a particular endocrine or neural system in the aging processes or the evidence that food restriction influences the functioning of a neural or endocrine system. Sapolsky et al (1986) have shown that rats exhibit a loss of regulatory control of glucocorticoid secretion with advancing age and suggest that hyperadrenocorticism is an important part of the aging process in this species. There is no information on the influence of food restriction on age changes in the regulation of glucocorticoid secretion and this should be rectified since it could be an important aspect of the action of food restriction. Another potential coupler is the glucose-insulin system. Cerami (1985) has hypothesized that glucose may act as a mediator of aging through the nonenzymatic reaction of glucose with proteins and nucleic acids to yield advanced glycosylation endproducts and thereby cause a loss of function. The extent of formation of advanced glycosylation endproducts increases with the concentration of glucose and the length of time exposed to that concentration of glucose. Unpublished work on the male Fischer 344 rats in our laboratory shows that food restriction significantly lowers plasma glucose concentration particularly during the dark phase of the light cycle and reduces the extent of glycation of hemoglobin. Other obvious potential couplers that should be explored are the thyroid because of the evidence that food restriciton decreases thyroid function (Wartofsky and Burman, 1982) and the sympathetic nervous system because of the hyperadrenergic state observed during aging (Rowe and Troen, 1980).

The recent findings in our laboratory of Kalu et al (1986) with Fischer 344 rats may have provided insight on how food restriction modulates age changes in endocrine glands. This study indicates that food restriction prevents the age-associated increase in the expression of the calcitonin gene in the thyroid C cell. Further work is needed to unequivocally establish this action for the thyroid C cell and to determine the extent to which such a mechanism applies to other endocrine cells.

Information is emerging on the influence of food restriction on age-changes in biochemical processes of most tissues and organs, presumably the targets of the endocrine and neural coupling. Richardson et al (1987) have found that the transcription of α_{2u}-globulin genes by isolated liver nuclei decreases with increasing age in male Fischer 344 rats supplied by our laboratory and that food restriction retards this age-related decline in gene expression. Simultaneously, in our laboratory Ward (1988) found that perfused livers from food restricted male Fischer 344 rats have a higher rate of protein synthesis over most of the life span than livers from ad libitum fed rats. Cheung and Richardson (1982) pointed out that maintaining protein turnover is important for cellular homeostasis and that functional deficits in aging may be due to a failure to do so. Clearly, influencing protein turnover could be a major way by which food restriction retards the aging processes and warrants further study.

Harman (1981) suggested that food restriction retards the aging process by reducing metabolic rate thereby decreasing free radical generation and damage. As discussed above, food restriction does not decrease metabolic rate. Nevertheless, work in our laboratory (Langaniere and Yu, 1987) indicates that food restriction may protect the male Fischer 344 rat from free radical damage. Food restriction was found to inhibit the age-related increase in malondialdehyde production by membranes isolated from the liver and also to reduce the age-related increase in lipid peroxide content of these membranes. It is further suggested that in part these findings result from food restriction inhibiting the age-related increase in membrane 22:5 fatty acid, thereby reducing the peroxidizability of membrane lipids. In this way, food restriction maintains membrane structure and fluidity. These provocative findings should be explored further.

REFERENCES

Berg BN, Simms HS (1960). Nutrition and longevity in the rat. II. Longevity and onset of disease with different levels of intake. J Nutr 71:255-263.

Bertrand HA, Anderson WR, Masoro EJ, Yu BP (1987). Action of food restriction on age-related changes in adipocyte lipolysis. J Gerontol 42:666-673.

Bertrand HA, Lynd, FT, Masoro EJ, Yu BP (1980a). Changes in adipose mass and cellularity through the adult life of rats fed ad libitum or a life prolonging restricted diet. J Gerontol 35:827-835.

Bertrand HA, Masoro EJ, Yu BP (1980b). Maintenance of glucagon promoted lipolysis in adipocytes by food restriction. Endocrin 107:591-595.

Brenner BM, Meyer TW, Hostetter TH (1982). Dietary protein intake and progressive nature of kidney disease. N Engl J Med 307:652-659.

Cerami A (1985). Hypothesis: Glucose as a mediator of aging. J Am Geriat Soc 33:626-634.

Cheung HT, Richardson A (1982). The relationship between age-related changes in gene expression, protein turnover and the responsiveness of an organism to stimuli. Life Sci 31:605-613.

Harman D (1981). The aging process. Proc Natl Acad Sci USA 78:7124-7128.

Herlihy JT, Yu BP (1980). Dietary manipulation of age-related decline in vascular smooth muscle functions. Am J Physiol 238:H652-H655.

Holehan AM, Merry BJ (1986). The experimental manipulation of ageing by diet. Biol Rev 61:329-368.

Iwasaki K, Gleiser CA, Masoro EJ, McMahan CA, Seo E, Yu BP (1988a). The influence of dietary protein source on longevity and age-related disease processes of Fischer rats. J Gerontol: Biol Sci 43:B5-B12.

Iwasaki K, Gleiser CA, Masoro EJ, McMahan CA, Seo E, Yu BP (1988b). Influence of the restriction of individual dietary components on longevity and age-related disease of Fischer rats: The fat component and the mineral component. J Gerontol: Biol Sci 43:B13-B21.

Kalu DN, Cockerham R, Yu BP, Ross BA (1983). Lifelong dietary modulation of calcitonin levels in rats. Endocrin 113:2010-2016.

Kalu DN, Hardin RR, Cockerham R, Yu BP, Norling BK, Egan J. (1984). Lifelong food restriction prevents senile osteoporosis and hyperparathyroidism in rats. Mech Ag Dev 26:103-112.

Kalu DK, Hardin RR, Yu BP, Kaplan G, Ferry S, Jacobs JW (1986). Effects of undernutrition on calcitonin and calcium metabolism in the rat. Eight Annual Sci Meet Am Soc Bone & Miner Res, Abst No 365.

Laganiere S, Yu BP (1987). Anti-lipoperoxidation action of food restriction. Biochem Biophys Res Comms 145:1185-1191.

Liepa GU, Masoro EJ, Bertrand HA, Yu BP (1980). Food restriction as a modulator of age-related changes in serum lipids. Am J Physiol 238:E253-E257.

Maeda H, Gleiser CA, Masoro EJ, Murata I, McMahan CA, Yu BP (1985). Nutritional influences on aging of Fischer 344 rats: II. Pathology. J Gerontol 40:671-688.

Masoro EJ (1985). Nutrition and aging - A current assessment. J Nutr 115:842-848.

Masoro EJ (1986). Physiology of aging. In Holm-Pedersen, Löe H (eds): "Geriatric Dentistry," Copenhagen: Munksgaard, pp 34-55.

Masoro EJ, Compton C, Yu BP, Bertrand H (1983). Temporal and compositional dietary restrictions modulate age-related changes in serum lipids. J Nutr 113:880-892.

Masoro EJ, Iwasaki K, Gleiser CA, McMahan CA, Seo E, Yu BP (1988). Comparison of dietary calories and proteins as factors in the life span progression of nephropathy in rats. FASEB J 2:A1208.

Masoro EJ, Yu BP, Bertrand HA (1982). Action of food restriction in delaying the aging process. Proc Natl Acad Sci USA 79:4239-4241.

McCarter R, Masoro EJ, Yu BP (1985). Does food restriction retard aging by reducing the metabolic rate? Am J Physiol 248:E488-E490.

McCay C, Crowell M, Maynard L (1935). The effect of retarded growth upon the length of life and upon ultimate size. J Nutr 10:63-79.

Richardson A, Butler JA, Rutherford MS, Semsei I, Gu M, Fernandes G, Chiang W (1987). Effect of age and dietary restriction on the expression of α_{2u}-globulin. J Biol Chem 262:12821-12825.

Rowe JW, Troen BR (1980). Sympathetic nervous system and aging in man. Endocrin Rev 1:167-178.

Sacher GA (1977). Life table modifications and life

prolongation. In Finch CE, Hayflick L (eds): "Handbook of the Biology of Aging," New York: Van Nostrand Reinhold, pp 582-638.

Sapolsky RM, Krey LV, McEwen BS (1986). The neuroendocrinology of stress and aging: The glucocorticoid cascade hypothesis. Endocrin Rev 7:284-301.

Walford RL (1985). The extension of maximum life span. Clin Geriatr Med 1:29-35.

Ward, WF (1988). Enhancement by food restriction of liver protein synthesis in the aging Fischer 344 rat. J Gerontol: Biol Sci 43:B50-B53.

Wartofsky L, Burman KD (1982). Alteration in thyroid function in patients with systemic illness: The "Euthyroid Sick Syndrome." Endocrin Rev 3:164-217.

Yu BP, Bertrand HA, Masoro EJ (1980). Nutrition - aging influence of catecholamine-promoted lipolysis. Metab 29:438-444.

Yu BP, Masoro EJ, McMahan CA (1985). Nutritional influences on aging of Fischer 344 rats: I. Physical, metabolic and longevity characteristics. J Gerontol 40:657-670.

Yu BP, Masoro EJ, Murata I, Bertrand HA, Lynd FT (1982) Life span study of SPF Fischer 344 male rats fed ad libitum or restricted diets: Longevity, growth, lean body mass and disease. J Gerontol 37:130-141.

LIFE SPAN AND PATHOLOGY

Dietary Restriction and Aging, pages 39-49
© 1989 Alan R. Liss, Inc.

THE DESIGN OF THE LOBUND AGING STUDY AND THE GROWTH AND SURVIVAL OF THE LOBUND-WISTAR RAT

David L. Snyder and Bernard S. Wostmann.

Lobund Laboratory, University of Notre Dame, Indiana, 46556.

The use of germfree (GF) animals in aging research was begun at Lobund Laboratory in 1958. The rationale for their use, as first expressed by Helmut Gordon, is that GF animals permit the distinction between deleterious effects of microbes on the aging host and the impairment in function due to endogenous and other non-microbial environmental agents (Gordon, 1959). In 1966 Gordon clearly showed that GF mice lived at least eight months longer than conventional (CV) mice (24 vs. 16) (Gordon et al., 1966a). The leading causes of death among the CV mice were respiratory infection (38%) and kidney lesions (14%), while GF mice died of intestinal atonia (36%), related to enlarged and twisted cecums, and kidney lesions (12%). Researchers at Lobund continued to examine aging in long-lived GF animals (Pollard, 1971) and to demonstrate the spontaneous development of liver tumors and benign adenomas of endocrine glands in GF rats older than 30 months (Pollard and Luckert, 1979). GF rats and mice have lower resting oxygen consumption, cardiac output, and reduced heart size (Wostmann, 1975) and reduced adult body size (Snyder and Wostmann, 1987) when compared to CV counterparts. These characteristics of the GF animal may contribute to their longer life span.

The effect of reduced dietary intake on longevity in GF rats was first reported in 1985 (Pollard and Wostmann, 1985). This pilot study used ten GF Lobund-Wistar (L-W) rats restricted to 12 grams of diet a day from weaning. The approximately 30% reduction in food intake from adult levels resulted in all ten rats living to 37 months of age and having none of the characteristic neoplasms of ad libitum fed GF L-W rats of that age. These findings eventually led to the much larger Lobund Aging Study which is described in these proceedings. This chapter will give details on care of the L-W rats, their growth and survival, the sacrifice methods used during this study, how tissues were stored and distributed, and age-related changes in selected organ weights.

METHODS

The L-W strain of rats originated at Notre Dame in 1958 with the creation of a GF breeding colony. The closed colony is now in its 56th generation. The CV breeding colony was derived from the GF colony, and at regular intervals GF males and females are added to the CV colony to maintain close genetic proximity. All L-W rats are free of pathogenic microorganisms including viruses. Nephrosis is not evident in the L-W rats until after 30 months of age and is never severe. Further details on age-related pathology in the L-W rat are provided in these proceedings in the chapter by Dr. Morris Pollard.

Only male rats were used in this study. The GF rats were housed in plastic and steel isolators and were maintained using routine gnotobiotic procedures (Subcommittee on Standards for Gnotobiotics, 1970). The CV rats were housed in plastic isolators which were open to the local environment for introducing feed and water, and the rats were weighed outside the isolator. All rats were weighed each week from age 6 to 10 weeks and thereafter once every 4 weeks. Full-fed (ad libitum fed) rats (GF-F and CV-F) were housed 4 to a cage (commercial plastic boxes) which measured approximately 18 x 9.5 x 8 inches. There were two cages per isolator. Restricted fed rats (GF-R and CV-R) were taken from the colony at 6 weeks of age and housed individually in cages measuring 14 x 8 x 7 inches. All rats were kept on Sani Cell corncob bedding and given untreated tap water. The rooms containing the isolators were air- and humidity-controlled with 12 hour light/dark cycles. All rats were fed steam-sterilized natural ingredient diet, L-485, our colony diet since 1968 (Kellogg and Wostmann, 1969). The ingredients and nutrient composition of L-485 are given in Table 1. Extra amino acids and vitamins are included in the diet to compensate for losses due to sterilization. The restricted rats were never allowed more than 12 g of diet per day. This method of feeding becomes restrictive at about 8 weeks of age and results in a 30% reduction in feed intake in adult rats (Snyder and Wostmann, 1987).

TABLE 1. Composition of natural ingredient diet L-485.

Ingredients	g/Kg	Nutrient Composition %	
Ground corn	590	Protein	20
Soybean meal, 50% CP[*]	300	Fat	5.3
Alfalfa meal, 17% CP	35	Fiber	3.0
Corn oil	30	Ash	5.5
Iodized NaCl	10	Moisture	11.2
Dicalcium phosphate	10	Nitrogen-free	
Calcium carbonate	5	extract	55
Lysine	5		
dl-Methione	5	Gross energy 3.9 Kcal/g	
Vitamin and mineral mix	.25		
BHT	.125		

[*] crude protein

Blood and tissue samples from eight healthy L-W rats at 6, 18, and 30 months of age from each of the four experimental groups were distributed to interested investigators. In addition to these rats, moribund rats older than 24 months were also distributed. A smaller group of investigators also received samples from L-W rats between 2 and 7 months of age. After an overnight fast, each rat was anesthetized with halothane and blood was removed from the exposed heart with a needle and syringe. Five ml of the blood were mixed with heparin and the remaining blood was allowed to clot for 30 minutes and then centrifuged. The serum was frozen at -70°C in individual aliquots for each investigator. Rats were examined for tumors and tissues were processed for histopathology. All rats were killed within 30 minutes of each other beginning at 9 A.M. No more than four rats were killed in one day. Individual tissues were quickly removed, weighed and frozen in liquid nitrogen or preserved according to the protocol of each investigator receiving tissue samples.

RESULTS

A survival distribution function (SAS Institute Inc., 1985) was calculated for each experimental group from 100 CV-F, 88 CV-R, 96 GF-F and 127 GF-R rats which died after 12 months of age. This includes healthy rats sacrificed at 18 and 30 months. The median survival age in months for each group was CV-F: 31.0, CV-R: 39.0, GF-F: 33.6, and GF-R: 37.8. Table 2 gives the percent survival of each group at selected ages. A Wilcoxon test of equality between each of the survival distribution functions showed that the survival of the restricted rats was different from that of the full-fed rats, that the survival of the GF-F rats was different from that of the CV-F rats, and that there was no difference in survival between GF-R and CV-R rats.

As shown in Table 3 each of the experimental groups had a characteristically different pattern of growth. Up to six months of age the GF-F rats grow at the same rate as the CV-F rats. After seven months the CV-F rats grow at a slightly faster rate. By 18 months the average weight of CV-F rats is just over 480 grams but the GF-F rats average just over 450 grams. Since the cecum of the GF rat is 20 grams larger than that of a CV rat the actual difference in body weights between GF-F and CV-F rats is closer to 50 grams. The CV-R and GF-R rats grew at a similar rate. The GF-R rats eventually outweighed the CV-R rats but after compensating for the 20 gram cecum the difference was only 10 grams. Further details on the growth of GF and CV L-W rats are available in Snyder and Wostmann (1987).

TABLE 2. Percent survival of male Lobund-Wistar rats at selected ages.

Age in Months	CV-F	CV-R	GF-F	GF-R
18	89	97	100	100
21	80	97	100	100
24	69	95	100	100
27	66	94	92	94
30	56	81	67	91
33	32	72	54	81
36	12	67	36	66
39	1	49	9	45
42	--	20	--	30

N: 100 CV-F, 88 CV-R, 96 GF-F, 127 GF-F

TABLE 3. Mean body weight in grams of male Lobund-Wistar rats.

Age in Months	CV-F	CV-R	GF-F	GF-R
2	185	170	210	173
6	355	252	340	258
12	443	290	411	306
18	483	298	448	321
24	475	295	448	329
30	450	285	425	331
36	---	283	380	320

Table 4 lists the body and organ weights of the rats sacrificed at 6, 18 and 30 months and then distributed to the collaborators in the Lobund Aging Study. Young and adult GF rats are known to have smaller livers and hearts when compared to CV rats (Gordon et al., 1966b; Wostmann et al., 1968). Data from our study confirm this relationship from 6 to 30 months of age both for absolute weight and as a percent of body weight. Liver and heart weight continued to increase with age in all groups. The R rats had higher liver and heart weights as a percent of body weight at all ages when compared to the F rats. Gastrocnemius muscle weight declined in the CV-F and GF-F rats between 6 and 30 months but not in the CV-R and GF-R rats. The epididymal fat pad increased in weight between 6 and 18 months in all groups, but then declined in weight between 18 and 30 months. As a percent of body weight the R rats had less epididymal fat pad and more gastrocnemius at each age when compared to the F rats. Restricting intake to 12 grams per day reduced body weight in the CV rats by 25% and GF rats by 29% at 6 months of age. At 18 and 30 months the reduction in body weight was near 35% in CV rats and near 24% in GF rats. The percent reduction in liver and heart weight due to restriction was less than the percent reduction in body weight. The percent reduction in epididymal fat pad weight was much higher. Restriction reduced fat pad size by 50 to 56% in the CV rats and by 45% in the GF rats.

TABLE 4. Body and organ weights of L-W rats in grams.*

	Body Weight	Liver	Heart	Muscle**	Fat***
6 months					
CV-F	374	9.9(2.7%)	1.3(.34%)	2.0(.53%)	5.0(1.3%)
CV-R	280	8.5(3.0%)	1.0(.36%)	1.5(.53%)	2.5(.9%)
GF-F	356	7.1(2.0%)	1.1(.30%)	1.8(.51%)	4.5(1.3%)
GF-R	252	6.4(2.5%)	0.8(.31%)	1.3(.51%)	2.5(1.0%)
18 months					
CV-F	458	11.1(2.4%)	1.6(.35%)	1.7(.38%)	7.5(1.6%)
CV-R	298	8.4(2.8%)	1.3(.44%)	1.5(.52%)	3.7(1.2%)
GF-F	405	8.4(2.1%)	1.2(.30%)	1.4(.34%)	6.3(1.5%)
GF-R	308	7.8(2.5%)	1.0(.32%)	1.4(.45%)	3.5(1.1%)
30 months					
CV-F	457	12.2(2.7%)	1.8(.40%)	1.5(.32%)	6.2(1.4%)
CV-R	306	8.9(2.9%)	1.3(.42%)	1.5(.48%)	2.7(.9%)
GF-F	415	9.9(2.4%)	1.4(.34%)	1.1(.27%)	4.7(1.1%)
GF-R	323	8.5(2.7%)	1.1(.33%)	1.4(.43%)	2.6(.8%)

*N=8; percent of body weight for each organ is in parentheses.
**Gastrocnemius muscle. ***Epididymal fat pad

Similar age-related changes in organ weights can be seen in the data from all the healthy rats sacrificed during the entire Lobund Aging Study. The relationship between organ weight and age was examined by linear regression in 76 CV-F, 48 CV-R, 81 GF-F and 74 GF-R rats between 6 and 30 months of age. Liver weight showed significant (P<.01) positive slopes and coefficients of determination (R^2) between .29 and .42 for each of the groups. The slope was higher in the F rats. Heart weight also showed significant positive slopes with R^2s between .42 and .57 for each of the groups. Gastrocnemius muscle weight showed a significant negative slope but only in the F rats. The R^2s were CV-F = .33 and GF-F = .41. There was no linear change with age in the epididymal fat pads.

DISCUSSION

The GF animal has been suggested as an ideal model for aging research because of the freedom from microbial interference during aging, but the cost of maintaining GF animals for extended periods is prohibitive. Our study has shown that CV rats derived from a clean breeding colony and housed in minimal barrier isolation will live almost as long as GF rats. The increase in percent survival of GF-F rats at all ages when compared to CV-F rats is due partly to the lack of prostate infections in GF rats. Approximately 25% of all deaths in CV-F rats were due to prostatitis and the majority of cases occurred between 18 and 24 months of age. Surprisingly, prostatitis occurred in only 8% of the CV-R rats with no relationship to age.

The reduced adult body weight of the GF-F rats when compared to the CV-F rats may have contributed to the greater 10% survival (39.0 vs. 36.3 months) and maximum survival (40.5 vs. 38.9 months) ages of GF-F rats. The possible influence of a natural dietary restriction extending the life span of the GF-F rats cannot be overlooked. This possibility is also suggested by the similarity in life spans between CV-R and GF-R rats which have the same feed intake and similar adult body weights. Our study suggests that there is a strong association between adult body weight and longevity within the L-W rat strain. The CV-F rats, being the heaviest group, had the shortest median survival age. The two restricted fed groups which had similar adult body weights but dramatically different physiological adaptations to their environments had nearly identical survival distribution functions. The GF-F rats which had an intermediate adult body weight had a median survival age between the CV-F and restricted fed groups.

It is unclear why the reduced metabolic rate of the GF rat, as indicated by the reduced liver and heart size in both GF-F and GF-R rats, did not influence life span. The reduction in heart size, cardiac output and resting oxygen consumption of GF animals has been related to bioactive compounds produced in the enlarged cecum (Wostmann et al., 1968). These conditions may be responsible in part for the lower adult body size of the GF-F rats, but were without effect when dietary intake was limited. A reduction in the generation of oxygen free radicals due to reduced oxygen consumption has been proposed as a possible mechanism for the effects of dietary restriction on aging and disease (Harman, 1986). Free radical generation as a major cause of aging is not well supported (Sohal, 1987) and reduced oxygen consumption seems to be unrelated to life extension in diet-restricted rats (Masoro, 1988). Our study with GF rats does not support the theory that reduced oxygen consumption extends life span. This does not rule out the involvement of the cellular antioxidant defense system and oxidative damage to proteins in the ability of dietary restriction to extend life span (see the chapters by Chen, Lang and Starke-Reed in these proceedings). Metabolic changes other than reduced oxygen consumption may contribute to life extension in diet-restricted rats. As a percent of body weight, the liver, heart, and gastrocnemius muscle were higher and the epididymal fat pad was lower in diet-restricted rats when compared to ad libitum fed rats. These differences indicate that specific metabolic adaptations have been made in the restricted fed rats.

Masoro (1980) has reported that in most of the rat strains used in aging research the median length of life is between 24 and 29 months and that diet and housing can have considerable influence on life span. The extended median length of life of the ad libitum fed L-W rat when compared to other rat strains (31 vs. 25 months) is primarily due to the lack of kidney disease in the L-W rat. Kidney disease is the major cause of death in most of the rat strains used in aging research. The development of this disease can be altered by dietary restriction (Maeda et al., 1985) or by the use of soy protein instead of casein as a protein source in the diet (Iwasaki et al., 1988). Nephropathy is a major problem in the use of rats for studying alterations in the aging process, especially when distinguishing age-related from drug-related effects during chronic toxicity studies (Goldstein et al., 1988). In the present study nephropathy was found in only 12% of CV-F and 5% of CV-R L-W rats. The condition was always mild and was never the cause of death. Pollard (1971) reported that during a 1961 examination 6 of 16 GF L-W rats over 24 months of age had nephritis, but that a similar examination in 1970 found no nephritis in 41 GF L-W rats over 24 months of age. The second group of rats was maintained on the natural ingredient diet L-485 which contains soybean meal as the major protein source. The earlier diet was semirefined and contained casein as the protein source.

The conventional Lobund-Wistar rat is an excellent model for examining the interaction of nutrition, tumor development and the aging process. The lack of kidney disease in the ad libitum fed L-W rat has allowed the striking effect of diet restriction on neoplastic disease to be seen (see the chapter by Pollard in these proceedings). True age-related changes can be monitored in the L-W rat between 18 and 30 months without interference from kidney disease or prominent neoplastic disease. Rats with prostate infections and prostate tumors can be eliminated from study by palpation of the lesion. In this way the ad libitum and restricted fed L-W rat may provide the standard or base rodent model that Masoro (1988) suggests is greatly needed for aging research.

REFERENCES

Goldstein RS, Tarloff JB, Hook JB (1988). Age-related nephropathy in laboratory rats. FASEB J 2:2241-2251.

Gordon HA (1959). The use of germ-free vertebrates in the study of "physiological" effects of the normal microbial flora. Gerontologia 3:104-114.

Gordon HA, Bruckner-Kardoss E, Staley TE, Wagner M, Wostmann BS (1966a). Characteristics of the germfree rat. Acta Anat 64:301-323.

Gordon HA, Bruckner-Kardoss E, Wostmann BS (1966b). Aging in germ-free mice: Life tables and lesions observed at natural death. J Gerontol 21:380-387.

Harman D (1986). Free radical theory of aging: role of free radicals in the origination and evolution of life, aging, and disease processes. In Johnson JE, Walford R, Harman D, Migueal J (eds): "Free Radicals, Aging, and Degenerative Disease." New York: Alan R. Liss, pp. 3-49.

Iwasaki K, Gleiser CA, Masoro EJ, McMahan CA, Seo E, Yu BP (1988). The influence of dietary protein source on longevity and age-related disease processes of Fischer rats. J Gerontol: Biol Sci 43:B5-B12.

Kellogg TF, Wostmann BS (1969). Stock diet for colony production of germfree rats and mice. Lab Anim Care 19:812-814.

Maeda H, Gleiser CA, Masoro EJ, Murata I, McMahan CA, Yu BP (1985). Nutritional influences on aging of Fischer 344 rats: II. Pathology. J Gerontol 40:671-688.

Masoro EJ (1980). Mortality and growth characteristics of rat strains commonly used in aging research. Exp Aging Res 3:219-233.

Masoro EJ (1988). Food restriction in rodents: an evaluation of its role in the study of aging. J Gerontol: Biol Sci 43:B59-B64.

Pollard M (1971). Senescence in germfree rats. Gerontologia 17:333-338.

Pollard M, Luckert PH (1979). Spontaneous liver tumors in aged germfree Wistar rats. Lab Anim Sci 29:74-77.

Pollard M, Wostmann BS (1985). Aging in germfree rats: the relationship to the environment, diseases of endogenous origin, and to dietary modification. In Archibald J, Ditchfield J, Rowsell HC (eds): "The Contribution of Laboratory Animal Science to the Welfare of Man and Animals," New York: Gustav Fischer Verlag, pp 181-186.

SAS Institute Inc. (1985). "SAS User's Guide: Statistics, Version 5 Edition." Cary, NC: SAS Institute Inc., pp 529-557.

Sohal RS (1987). The free radical theory of aging: a critique. Rev Biol Res Aging 3: 431-449.

Snyder DL, Wostmann BS (1987). Growth rate of male germfree Wistar rats fed ad libitum or restricted natural ingredient diet. Lab Anim Sci 37:320-325.

Subcommittee on Standards for Gnotobiotics, Committee on Standards, Institute of Laboratory Animal Resources, National Resource Council (1970). "Gnotobiotes, Standards and Guide Lines for Breeding, Care and Management of Laboratory Animals." Washington, DC: National Academy of Sciences.

Wostmann BS, Bruckner-Kardoss E, Knight PL (1968). Cecal enlargement, cardiac output and O2 consumption in germfree rats. Proc Soc Exp Biol Med 128:137-141.

Wostmann BS (1975). Nutrition and metabolism of the germfree mammal. World Rev Nutr Diet 22:40-92.

SPONTANEOUS DISEASES IN AGING LOBUND-WISTAR RATS

Morris Pollard & Phyllis H. Luckert

Lobund Laboratory, University of Notre Dame,
Notre Dame, Indiana 46556

Among the important goals of investigations on laboratory animals are those that address their use in risk assessments of drugs and other chemicals, studies on physiological mechanisms, and as models for specific diseases affecting humans. The most important focus of activities should be the development of healthy animals. As with laboratory instrumentation, the use of defective animals will often yield misinformation. Test animals that manifest high levels of spontaneous diseases, possibly unrelated to the test materials, result in enormous wastes of time, money, and efforts. Since infectious diseases are for the most part under control, current problems with laboratory animals are the control of so-called endogenous diseases, exemplified by nephropathy and neoplasms. The results of the recently-completed investigations on microbial flora, diet and longevity demonstrate that dietary considerations are of very high significance in this triad of information.

Increasing evidence supports the view that full-fed obese laboratory rats are not "normal" because of the high incidence of "spontaneous" diseases among them. The supportive documentation has been accumulating since 1935 (McCay et al, 1935). From data presented in the present investigation, a moderate dietary restriction (30%), is associated with a lower level of spontaneous diseases and a prolongation of life-span; and this appears to include diseases that are controlled by genetic mechanisms.

Doubtless, this will contribute substantially to the improvement of experimental protocols for risk assessments of drugs and chemicals, and for the refinement of model disease systems, thereby excluding diseases that serve to complicate and to confuse the interpretation of results.

This Symposium on aging presents a multidisciplinary investigation of the male Lobund-Wistar (L-W) rat under controlled conditions of microbial flora and caloric consumption. It involved germfree (GF) and conventional (CV) L-W rats. The unique aspects of this inquiry provide that the broad spectrum of examinations reported in this Symposium were derived from a single strain of rat (L-W), of one sex (male), fed the same diet (L-485), under the same conditions in a single laboratory. All results of the tests that have been conducted are thus enhanced by the uniformity of protocol conditions under which each animal was examined.

This chapter of the Proceedings presents the results of gross and microscopic examinations of aging male L-W rats, as influenced by diet and microbial status. The experimental protocol is described by Snyder and Wostmann.

METHODS

Rats

GF L-W rats were randomly-propagated in Lobund Laboratory through 56 generations. At each generation level, breeding groups were conventionalized in clean isolated air-conditioned rooms in which they were further propagated. The CV rats were housed in plastic boxes on ground corn cob bedding and fed, ad libitum, diet L-485 (Kellogg and Wostmann, 1969) and tap water.

While L-W rats are not considered "inbred," they do accept reciprocal skin transplants. The GF rats were examined extensively for microbial flora and none was found. The CV L-W rats were free of detectable pathogens; and they have been free of pneumonia and nephropathy. A profile of age-related spontaneous diseases in full-fed GF L-W rats listed relatively high incidences of neoplastic changes in endocrine-and endocrine-regulated glands, and in their livers (Pollard and Luckert, 1979). About 10% of the GF rats (older than 30 months), developed large metastasizing prostate adenocarcinomas (Pollard, 1973; Pollard and Luckert, 1975). In the current project prostate tumors were detected at earlier ages by abdominal palpation, thereby reducing average age incidence below 30 months.

Investigations on the effects of calorie-restriction on male L-W rats were based on an original protocol designed by Wostmann in which weanling rats were fed 12 grams of L-485 diet/day (Pollard and Wostmann, 1985). This was calculated to be 25-30% below that consumed by full-fed rats. The diet-restricted rats were caged individually; and the full-fed rats were housed 4 per cage.

As described in Chapter 4 of this publication, 400 weanling male L-W rats were assigned to the aging project; they included 200 GF and 200 CV rats. Half of each group were fed, ad libitum, the steam-sterilized diet (L-485) and tap water, and the other half were fed, from weanling age, the same diet (L-485) reduced by 30% (i.e. 12 gm/day). At intervals, groups of rats in each category were selected arbitrarily for autopsy examinations. Rats that appeared sick were subjected to the same autopsy examinations. Each rat was weighed, anesthetized by **halothane and exsanguinated** from the exposed heart. Serums were stored at -70°C for future examinations. Organs were weighed and samples thereof were fixed in Bouin's solution for 18 hours, changed to 70% ethanol, and then processed for histological examinations of hematoxylin and eosin-stained tissue sections.

RESULTS

The diet-restricted rats were in excellent physical condition for most of their life-span. They were alert, very active, and weighed average 30% less than the ad libitum-fed counterpart rats. The rats that were examined prior to age 20 months were generally free of pathological changes. Rats that had died some time prior to examination were of limited use because of degenerative changes in their tissues; or lost through cannibalism among the full-fed rats.

Most of the pathological changes occurred in rats older than 20 months: they involved endocrine and endocrine-regulated glands and livers. The gross and microscopic lesions were similar to those that developed spontaneously in ad libitum-fed aged GF L-W rats (Pollard and Luckert, 1979). In general, incidences of diseases were lower and time-deferred among the diet-restricted rats.

Significant levels of diseases were manifested in 3 organ systems among all of the full-fed rats:

A. <u>Liver tumors</u>. The small white foci observed in many of the livers were actually aggregations of large hepatocytes with clear cytoplasms. These were interpreted as "storage lesions," and thus they were not listed as neoplasms.

As noted in Table 1, the incidence of hepatomas, especially among the diet-restricted CV rats, was lower (26%) than among the full-fed rats (57%). Hepatomas were detected at avg 31.3 months of age among the full-fed CV rats compared to avg 41 months among the CV diet-restricted rats. The hepatomas were usually visible and projected above the surfaces of the organs, and consisted of aggregations of large hepatocytes. The "normal" lobular pattern

TABLE 1. Diet-related Incidence of 3 Tumors in Male Conventional (CV) and Germ-free (GF) L-W Rats*

	Hepatoma (CA)	Tumors PA	Adrenal Medulla
CV Fullfed	38/66-57% (28%) 31.3**	13/66-19.6% 26.6	46/66-69.2% >19
CV Restricted	12/46-26% (25%) 41	2/46-4.3% 36.7	23/42-54.7% >20
GF Fullfed	18/40-45% (22%) 32.2	2/40-5% 26	25/40-62.5% >20
GF Restricted	22/70-31% (32%) 36.5	7/70-10% 36.7	34/67-50.7% >20

*Lobund-Wistar rats; Age range 20-40 months.
(CA) = Carcinoma. Underlined data are significantly different.
**Average incubation period in months.

was distorted and the edges of the benign tumor produced flat pressure patterns against the neighboring normal parenchyma. Mitotic figures were not detected among the hepatoma cells.

Hepatocarcinomas had developed among 25% and 28% of those CV rats with liver tumors that had been diet-restricted and full-fed respectively. The carcinomas were characterized by extensive distortion and disorganization of lobular patterns; the tumor cells were large with prominent nuclei, and usually with ground-glass appearing cytoplasms. Mitotic figures were observed frequently among them. None of the carcinomas had developed detectable metastatic lesions.

The effects of dietary-restriction on hepatoma development were not as decisive among GF rats as among the CV rats relative to incidence and average latent periods (Table 1). However, the histological patterns of all liver tumors in the GF rats were similar to those observed among the CV counterpart rats.

B. <u>Prostate diseases</u>. Three age-related prostate diseases were noted among the CV full-fed L-W rats: between ages 10 to 20 months, 20% had developed prostatitis; between ages 20 to 30 months 19.6% had developed prostate adenocarcinomas (PAs); and between ages 30 to 41 months, 55% had developed lesions of benign stromal hyperplasia (designated BPH).

<u>In rats with prostatitis</u>, the gland was palpable through the abdominal wall as a smooth-surfaced enlargement in the pelvic region. <u>Prostatitis</u> was not observed among the GF rats; and the incidence of this disease was lower among the CV diet-restricted rats than among the full-fed CV rats.

<u>The PAs were localized</u> initially in a dorso-lateral lobe of the gland, from which the tumorigenic process expanded into the entire gland, including most of the seminal vesicles. The disease-free prostate gland (less seminal vesicles) weighed avg 1.5 grams, and the PA-affected glands ranged in weights from 5 to 30 grams. The palpable PAs (> 0.5 cm in diameter) were hard with rough surfaces, and they were scirrhous when incised. The PAs were assessed histologically as moderately differentiated adenocarcinomas,

in which there were sheets of large epithelium-type cells which showed a tendency for glandular patterns (Pollard, 1973; Pollard and Luckert, 1975; Pollard, 1977). The very large PAs were necrotic in the central regions. In many areas of the intact tumor (in GF and in CV rats), there was an infiltration of neutrophilic leukocytes. The urinary obstruction resulting from the enlarged prostate caused markedly dilated urine-filled bladders, dilation of ureters and pressure atrophy of the medullary regions of the kidneys. At examination time, 90% of the rats with large PAs manifested two patterns of metastatic tumor spread: (a) PA cells had penetrated the capsule of the prostate gland thereby producing masses of so-called "pearls" (small round solid tumors), that were attached to the peritoneal surfaces including the surfaces of the visceral organs. (b) PA cells had spread through abdominal lymphatic channels to the lungs on and in which they produced visible foci of tumor cells similar to those that have been described (Pollard, 1973; Pollard and Luckert, 1975). There were individual rats in which both patterns of metastasis were observed. The tumorigenic process occurred among rats in the 4 noted categories, but the incidence of PAs were significantly reduced among the diet-restricted CV rats, from 19.6% to 4.3%. The average age at which PAs were detected was significantly deferred among the rats on the restricted diet: from 26.6 months to 36.7 months (Table 1). The incidence of PAs was higher among full-fed CV rats than among full-fed GF rats.

A third prostate disease, characterized by stromal hyperplasia, and (for convenience) referred to as benign prostatic hyperplasia (BPH), was observed rarely in rats under 30 months of age; however, among rats aged 30 to 41 months, 55% had developed the BPH lesion; and this lesion appeared among full-fed and diet-restricted GF and CV rats. Among the rats with BPH, there was a weight reduction of their prostate glands. The BPH lesion was characterized by marked proliferation of stromal elements (connective tissue and smooth muscle). The stromal hyperplasia was most prominently displayed in the central peri-urethral region in which the glands and ducts were distorted, constricted, and atrophied; but the glands in the peripheral regions of the gland were dilated. However, when the stromal hyperplasia process had extended into the peripheral regions of the prostate glands, the ducts and acini were distorted, constricted and atrophied by the surrounding nodular proliferation of stromal elements. In some of the constricted acini, multiple layers of epithelial cells could be conceivably confused with a neoplastic process.

Rats with PA showed no evidence of BPH, and vice versa.

Serum levels of testosterone in the aging L-W rats declined as the ages extended beyond 30 months.

C. <u>Adrenal Medullary Tumors</u>. The development of the proliferative medullary lesions in the adrenal glands was not influenced by dietary restriction, nor by microbial status of the host (Table 1). This lesion was rarely observed among rats under age 19 months. It usually resulted in 3 to 5X enlargement of the gland. The medullary lesion consisted of aggregations of hyperchromatic cells that usually filled the medullary region, which eventually resulted in pressure atrophy of the cortex. Mitotic figures were observed very rarely among the medullary cells.

It is speculated that the medullary lesion may actually represent a hyperplastic process, and not an actual neoplasm. There was no histological evidence of hypertension (thick arterial walls in the kidney and pancreas); and actual examinations of the rats for blood pressure levels were negative for hypertension.

D. <u>Lesions were observed in other tissues</u>, in low incidences, and predominantly in endocrine and endocrine-regulated glands. As exemplified in Table 2, the hyperplastic or benign neoplastic lesions were noted in lung, thymus, parathyroid and mammary glands; and in lesser numbers in the thyroid and pancreas glands. The incidences of the lesions were lower among the diet-restricted rats than among the full-fed rats.

DISCUSSION AND SUMMARY

Moderate diet restriction is a unique procedure for reducing the incidence and for deferring the ages of onset of spontaneous diseases; and for significant extension of life-span. This phenomenon has been described repeatedly during the past 50 years, but the exact mechanism(s) thereof are not known. Earlier investigations involved extreme reductions of dietary intake, almost starvation levels, from which death rates were relatively high. Later work

TABLE 2. Diet-related Tumors in Aged Germfree and Conventional L-W Rats

Neoplasm	Conventional (%) Full-fed	Restricted	Germfree (%) Full-fed	Restricted
Thymoma	6 (8.9)	2 (4.3)	3 (9.6)	0
Breast Adeno-fibroma	5 (7.4)	2 (4.3)	8 (25)	4 (6)
Lung Adenoma	4 (5.9)	1 (2.1)	4 (12.9)	7 (10.6)
Parathyroid Adenoma	4 (5.9)	1 (2.1)	1 (3.2)	1 (1.5)
# Total/# at risk	19/67 (28.3)	6/46 (13)	16/31 (51.6)	12/66 (18.1)

involved, for example, 60% reduction in dietary intake (6 gm/day), which deficit appeared to be extreme (Ross and Bras, 1971). Evidently, it is not necessary to resort to such extremes in dietary restriction to produce the benefits of that practice. It is fortuitous that 12 gms of feed/day resulted in healthy L-W rats, which fulfilled all of the expectations of benefits. This was in sharp contrast to the "abnormal" high incidence of spontaneous diseases among full-fed counterpart rats. In this respect, full-fed rats can be judged abnormal; and the same relationship was manifested in other strains of rats: nephropathy and neoplasms were the major causes of disease.

The investigations reported on here are unique because the experimental protocol was based on a single set of conditions relative to one animal strain (L-W), males, on the same diet, and defined environmental conditions (conventional and germfree). The disease profile on spontaneous and on induced diseases in the L-W rats was assembled during the past 25 years. The L-W rats are unique in manifesting a very low incidence of spontaneous nephropathy. They are unique in that (a) 19.6% develop metastasizing prostate adenocarcinomas spontaneously in average 26 months; and (b) 57% develop liver tumors in average 31 months. While the role(s) of diet L-485 in disease manifestations should be further investigated, this whole-grain diet has merit for investigations on longevity.

The 30% restriction in dietary intake that was initiated at weanling age, resulted in (a) a high quality animal; (b) a reduced incidence of prostate cancer from 19.6% to 4.3% with extension of average latent period from 26.6 months to 36.7 months; (c) a reduced incidence of liver tumors from 57% to 26% with extension of average latent period from 31.3 months to 41 months (Table 1); and (d) reduced incidences of a variety of other specific tumors (Table 2). It is significant to note that proliferative lesions in the medullary areas of the adrenal glands were not influenced by dietary restriction, nor by germfree status; and that the medullary lesion was not associated with detectable disease.

The stromal proliferative lesion in the prostates of aged rats may be a counterpart of benign prostate hyperplasia in man: it occurred in aged animals with reduced levels of testosterone, and predominantly in the periurethral region of the gland. However, there was no marked enlargement of the prostate gland. It is of significance to note that stromal hyperplasia and prostate adenocarcinoma were not observed in the same animals, and that the age-relationships of the two diseases were different.

REFERENCES

Kellogg TF, Wostmann BS (1969). Stock diet for colony production of germfree rats and mice. Lab Animal Care 19:812-814.

McCay CM, Crowell MF, Maynard LA (1935). The effect of retarded growth upon the length of life span and upon the ultimate body size. J Nutr 10:63-79.

Pollard M (1973). Spontaneous prostate adenocarcinomas in aged germfree Wistar rats. J Natl Cancer Inst 51:1235-1241.

Pollard M (1977). Animal model of human disease: Metastatic adenocarcinoma of the prostate gland. Am J Path 86:277-280.
Pollard M, Luckert PH (1975). Transplantable metastasizing adenocarcinomas in rats. J Natl Cancer Inst 54:643-649.
Pollard M, Luckert PH (1979). Spontaneous liver tumors in aged germfree Wistar rats. Lab Anim Sci 29:74-77.
Pollard M, Wostmann BS (1985). Increased life span among germfree rats. In Wostmann BS, Pleasants JR, Pollard M, Teah BA, Wagner M (eds):"Germfree Research: Microflora Control and its Application to the Biomedical Sciences," New York: Alan R. Liss, pp 75-76.
Ross MH, Bras G (1971). Lasting influence of early caloric restriction on prevalence of neoplasms in the rat. J Natl Cancer Inst 47:1095-1113.

EFFECTS OF DIETARY RESTRICTION ON BODY COMPOSITION AND BODY SIZE IN GERMFREE AND CONVENTIONAL LOBUND-WISTAR RATS

G.A. Boissonneault, T. Giles, and P. B. Meyers

University of Kentucky, Department of Clinical Nutrition, Lexington, Kentucky 40536

INTRODUCTION

As early as 1914 calorie restriction was reported to depress spontaneous tumor formation in mice (Rous). This observation has been investigated with enthusiasm since the late 1930's, with the names of Sivertsen (1938), Tannenbaum (1940, 1942, 1944, 1945a, 1945b, 1949), Boutwell (1949), Rusch (1944, 1945, Boutwell et al., 1949), and Baumann (Lavik and Baumann, 1943; Rusch et al., 1945) standing out among the long list.

During this same time period, calorie restriction was described as an effective means for the prolongation of life span (McCay et al., 1935, 1939, 1943; Silverberg and Silverberg, 1955a, 1955b). It has been suggested to achieve this effect by slowing the processes of aging (Masoro et al., 1980). In 1984, Masoro reviewed four major hypotheses put forward to explain life extension by calorie restriction. These include delaying maturation, slowing the rate and duration of growth, reducing body fat, and diminishing the metabolic body rate per unit body mass. In this report, Masoro suggested a fifth hypothesis, that calorie restriction increased lifespan by acting on a "specific metabolic process", which itself influences physiologic and immunologic deterioration, the ultimate director of aging as we know it.

This study sought to investigate a sixth hypothesis, that a combination of body composition and body size are related to longevity and risk for diseases of aging. In fact, this and the other hypotheses mentioned above are most probably interdependent, and to some extent true. The composition of carcasses from conventional and gnotobiotic Lobund-Wistar rats, either fed *ad libitum* or restricted to approximately 70% of *ad libitum* consumption, was determined. These data are presented at this time. Correlations of body composition and body size with longevity and disease resistance will be reported at a later date.

METHODS

Animals

Male Lobund-Wistar rats from the breeding colony of the Lobund Laboratory, Notre Dame University, South Bend, Indiana, were used in these studies. Animals were raised in an open air or in a germ-free environment. A detailed description of housing and feeding conditions can be found in the report by Dr. Snyder in this symposium. Animals ranging from 6 months to 45 months of age were used in this study, and data are grouped by housing method (conventional or germ-free), feeding regimen (full-fed or restricted feeding), and age at death.

Body composition analysis

The analyses presented in this report were conducted on headless, exsanguinated, eviscerated carcasses of rats which had been frozen following removal of tissues for analysis by other investigators. Thus, in this paper, carcass composition will be defined as the percent of the eviscerated, exsanguinated, and headless carcass as protein, fat, water, and ash.

Frozen carcasses were thawed, weighed, cut into 2-3 cm cubes, and placed into 1 quart canning jars. Approximately 1.5 x of the carcass weight in distilled water was added to each jar and all weights recorded. The jars were loosely covered and autoclaved for 1 hour at 15 pounds pressure, removed from the autoclave, and allowed to cool at room temperature. When cooled, the contents of the jars were placed into a large stainless steel Wharing blender jar and homogenized at high speed for 10 minutes. After homogenization, aliquots weighing approximately 100 grams were removed while constantly stirring the mixture. Aliquots were removed to labeled plastic whirl-pack sacks and frozen until analyzed.

Weighed aliquots were placed into tared 250 ml beakers and dried for 24-36 hours or to constant weight in a drying oven held at 105°C. After cooling beakers were weighed and dry weight determined. From this value and the prerecorded weight of the carcass, the amount of water in the original carcass was calculated. Dried material was blended in a small stainless steel blender, resulting in a finely divided composite of the original carcass, minus water, and weighed aliquots were removed for the determination of total ash and fat.

Carcass ash was determined by placing weighed aliquots of the dry material into tared crucibles and incinerating in a muffle furnace for 12-16 hours at 500°C. When cool crucibles were again weighed and ash weighed calculated. From these data total carcass ash content was calculated. To determine carcass fat weighed aliquots were placed into tared filter paper envelopes and repeatedly extracted for 8-12 hours with diethyl ether in a modified Soxlet apparatus. When dry, envelopes were again weighed and fat determined as weight lost by extraction. Once again, total carcass fat content was calculated from these data. Carcass protein content was calculated as carcass weight - (fat + ash + water).

Carcass composition data are reported as percent carcass composition of individual components. Carcass size data are expressed as lean body mass (LBM - fat-free carcass weight) and fat mass (FM - total carcass fat).

Data are presented as mean ± standard deviation, with sample size ranging from 3 to 12 rats/time point/treatment group.

RESULTS

Both full-fed groups, i.e., conventional full-fed (CF) and germ-free full-fed (GF), weighed more than their food-restricted counterparts (conventional restricted - CR, and germ-free restricted GR) throughout the experiment (Figure 1A). Moreover, germ-free groups tended to weigh less than conventionally raised rats, although this difference was most noticeable in the full-fed groups. In all groups carcass weight rose to a peak at 26-32 months of age, after which it began to decline and reached a new low by 37-39 months of age. Analysis of the carcass composition (Figure 1B, C, and D) demonstrate few differences in body

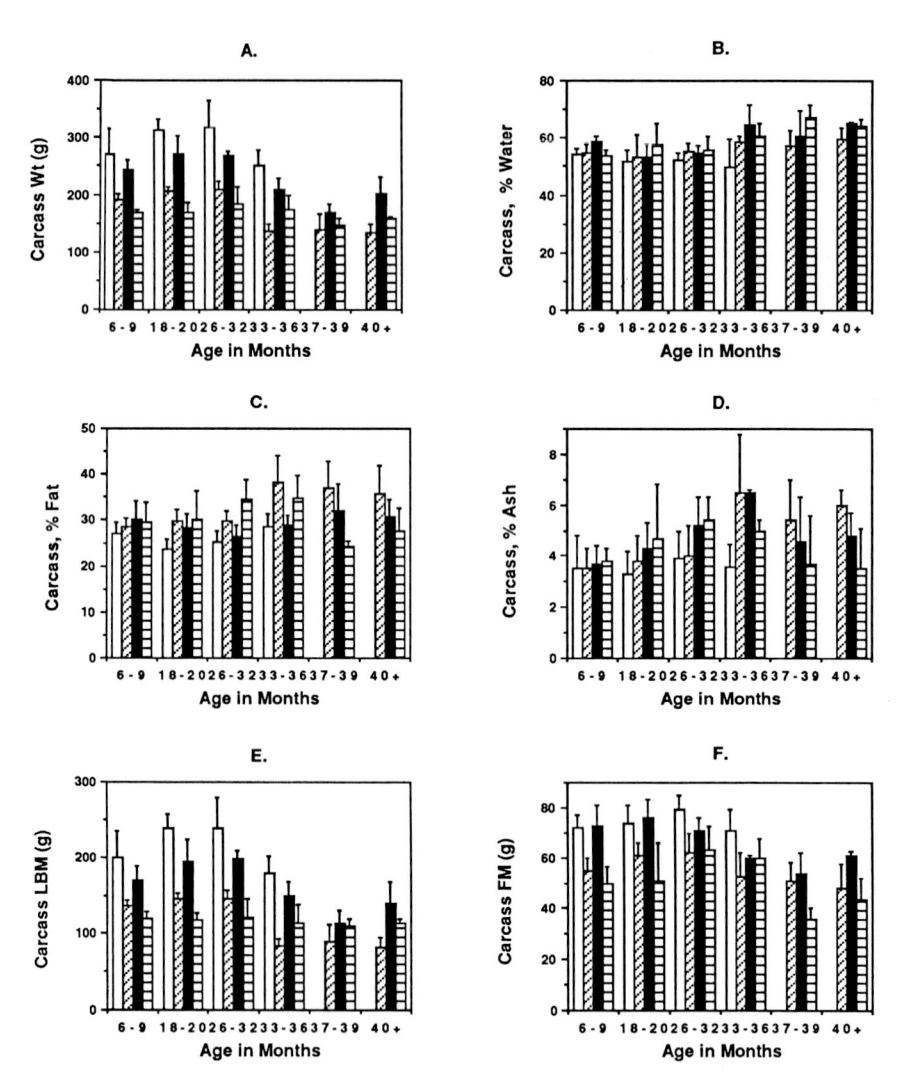

Figure 1: Relationship of environment (conventional - C, germ-free - G), feeding schedule (full-fed - F, restricted - R), and age in months on A) carcass weight, B) carcass water content, C) carcass fat content, D) carcass ash content, E) carcass lean body mass (LBM), and F) carcass fat mass (FM). Bars indicate mean ± S.D., n = 3 to 12 rats. ☐ CF ▨ CR ■ GF ☷ GR

composition, regardless of environmental conditions or feeding regimen.

Thus, little if any difference in body composition was noted among treatment groups, although body size (indicated by carcass weight) was greater in full-fed than in restricted, and in conventionally raised than in germ-free. Since body composition exhibited little variation it is not surprising that the lean body mass (LBM) and fat mass (FM) components of body size vary directly with carcass weight (Figure 1E and F, respectively). Full-fed groups tended to have more carcass fat than did restricted animals. LBM was also greater in the full-fed groups and remained so throughout the experiment. Conventionally raised rats tended to have a larger LBM than their germ-free counterparts. The loss of carcass weight after 26-32 months of age was proportionally equal from LBM and FM components.

DISCUSSION

In this experiment, food restriction did not result in changes of body composition, but only of body size. A similar outcome was reported by Boissonneault et al. (1986). In this study, one group of rats was fed a semipurified diet containing 30% fat but restricted to 84% of the calories consumed by ad libitum 30% fat-fed rats. The calorie restricted rats exhibited the same body composition as their ad libitum-fed pairs, but their body size, whether measured as LBM, FM, or carcass weight, was reduced. While body composition was unchanged in the restricted rats relative to their full-fed counterparts, the incidence of 7,12-dimethylbenz[a]anthracene-induced mammary tumors was greatly depressed. These results suggest that the risk for tumors in the present experimental groups may be related to lean body mass, which did vary significantly. This relationship is currently being investigated.

Because the manner of sample preparation resulted in carcasses without viscera, blood, or heads, data are probably skewed to at least some extent, e.g., abdominal fat was not available for analysis, and without this major fat depot the measured carcass fat mass may not reflect that present in the carcass at the time of death. This makes estimation of true body composition somewhat difficult, although the study by Boissonneault et al. (1986) using calorie restricted and full-fed rats referred to above suggests that body composition may in fact not have been vastly changed by the dietary manipulations used in this experiment.

Epidemiological studies support a relationship of body weight/height on morbidity and mortality (Build Study, 1979), and body weight and height on risk for cancer (Micozzi, 1985). Each of these measurements serve as indices of body composition and body size, with body weight/height serving as an index of relative obesity and body height reflecting lean body mass. Studies with experimental animals indicate that the contribution of lean body mass or body weight to longevity (Lesser et al., 1973; Yu et al., 1982; Beauchene et al., 1986) and cancer risk (Boissonneault et al., 1986; Albanes, 1987) is perhaps more important than that of obesity per se, although obesity itself may impart some effect on cancer risk (Boissonneault et al., 1986).

Other participants of the Lobund symposium on aging and dietary restriction in conventional and germ-free animals will discuss the effects of these variables on longevity and disease, including cancer.

REFERENCES

Albanes D (1987). Total calories, body weight, and tumor incidence in mice. Cancer Res 47:1987-1992.

Beauchene RE, Bales CW, Bragg CS, Hawkins ST, Mason RL (1986). Effect of age of initiation of feed restriction on growth, body composition, and longevity of rats. J Gerontol 41:13-19.

Boissonneault GA, Elson CE, Pariza MP (1986). Net energy effects of dietary fat on chemically induced mammary carcinogenesis in F344 rats. J Natl Cancer Inst 76:335-338.

Boutwell RK, Brush MK, Rusch HP (1949). The stimulating effect of dietary fat on carcinogenesis. Cancer Res 9:741-746.

Build Study, 1979. Society of Actuaries and Association of Life Insurance Medical Directors of America. Recording and Statistical Corporation, 1980.

Lavik PS, Baumann CA (1941). Further studies on tumor-promoting action of fat. Cancer Res 3:749-756.

Lesser GT, Deutsch S, Markofsky J (1973). Aging in the rat: Longitudinal and cross-sectional studies of body composition. Am J Physiol 225:1472-1478.

Masoro EJ, Yu BP, Bertrand HA, Lynd FT (1980). Nutritional probe of the aging process. Fed Proc 39:3178-3182.

Masoro EL (1984). Food restriction and the aging process. J Am Geriat Soc 32:296-300.

McCay CM, Crowell MF, Maynard LM (1935). The effect of retarded growth upon the length of life span and upon the ultimate body size. J Nutr 10:63-79.

McCay C, Maynard L, Sperling G, Barnes L (1939). Retarded growth, life span, ultimate body size and age changes in the albino rat after feeding diets restricted in calories. J Nutr 18:1-13.

McCay C, Sperling G, Barnes L (1943). Growth, ageing, chronic diseases, and life span of rats. Arch Biochem 2:469-479.

Micozzi MS (1985). Nutrition, body size, and breast cancer. Yearbook of Physical Anthropology 28:175-206.

Rous P (1914). The influence of diet on transplanted and spontaneous tumors. J Exp Med 20:433-451.

Rusch HP (1944). Extrinsic factors that influence carcinogenesis. Physiol Rev 24:177-204.

Rusch HP, Kline BE, Baumann CA (1945). The influence of caloric restriction and of dietary fat on tumor formation with ultraviolet radiation. Cancer Res 5:431-435.

Silverberg M, Silverberg R (1955a). Diet and life span. Physiol Rev 35:347-362.

Silverberg M, Silverberg R (1955b). LIfe span of "yellow" mice fed enriched diets. Am J Physiol 181:128-130.

Silvertsen I, Hastings WH (1938). A preliminary report on the influence of food and function on the incidence of a mammary gland tumor in "A" stock albino mice. Minnesota Med 21:873-875.

Tannenbaum A (1940). The initiation and growth of tumors. Introduction. I. Effects of underfeeding. Am J Cancer 38:335-350.

Tannenbaum A (1942). The genesis and growth of tumors II. Effects of caloric restriction *per se*. Cancer Res 2:460-467.

Tannenbaum A (1944). The dependence of the genesis of induced skin tumors on the caloric intake during different stages of carcinogenesis. Cancer Res 4:673-677.

Tannenbaum A (1945a). The dependence of tumor formation on the degree of caloric restriction. Cancer Res 5:609-615.

Tannenbaum A (1945b). The dependence of tumor formation on the composition of the calorie restricted diet as well as on the degree of restriction. Cancer Res 5:616-625.

Tannenbaum A, Silverstone H (1949). The influence of degree of caloric restriction on the formation of skin tumors and hepatomas in mice. Cancer Res 9:724-727.

Yu BP, Masoro EJ, Murata I, Bertrand HA, Lynd FT (1982). Life span study of SPF Fischer 344 male rats fed *ad libitum* or restricted diets: Longevity, growth, lean body mass and disease. J Gerontol 37:140-141.

CARDIAC FIBROSIS IN THE AGED GERMFREE AND CONVENTIONAL LOBUND-WISTAR RAT

Gibbons G. Cornwell III and Beverly P. Thomas

Department of Medicine, Dartmouth Medical School, Hanover, New Hampshire 03756

INTRODUCTION

In view of the prolonged survival of Lobund-Wistar rats maintained on low caloric intake, a morphologic study of heart tissues from elderly germfree and conventional rats was undertaken for the presence or absence of age-related cardiomyopathy. Since two specific types of senile cardiac amyloid are found in man (Westermark et al., 1979; Westermark et al., 1977), stains for amyloid were performed initially. Results showed no evidence of amyloid, but collagen deposition was markedly increased in the hearts of old rats. The degree of fibrosis was significantly reduced in animals fed a restricted diet.

METHODS

A total of 127 rats were studied. Eight rats were studied from each of the four groups: conventional fullfed (CV-F), conventional restricted diet (CV-R), germfree fullfed (GF-F) and germfree restricted diet (GF-R) at 7 mo, 18 mo and 30 mo. Thirty-one rats were studied in the 32-48 mo age group. Rats were anesthetized with halothane and tissues frozen at -70C. Hearts, kidneys, livers and lungs were shipped on dry ice from the Lobund Laboratories to Dartmouth. Whole hearts were cut sagittally (to allow study of atria and ventricles), processed by the Sainte-Marie method (Sainte-Marie, 1962) and imbedded in paraffin.

Six micra sections were stained with alkaline Congo red (Puchtler and Sweat, 1965) Gomori's reticulum stain (Gomori, 1973) and Masson trichrome stain (Masson, 1929; Lillie, 1985). Congo red stains were examined for amyloid with polarized white light and green filtered UV light. Anti-P component was used to examine tissues containing equivocal amyloid deposits (Sternberg, 1979; Shirahama et al., 1981). Reticulin and collagen depositions were graded from 0 (absent) to 4 (extensive). All tissues were examined blindly by two investigators.

Statistical analysis was performed using the F Test-Analysis of Variance (Cochran and Snedecor, 1980), Dunnett's Test (Winer, 1971) and the Student t Test (Cochran and Snedecor, 1980).

RESULTS

No amyloid deposits were present in 240 tissues studied (49 atria, 144 ventricles, 20 lungs, 27 livers). Most of these tissues (68%) were derived from elderly rats (30-48 mo).

Collagen deposition in heart tissues increased with age. In CV-F animals, the mean degree of ventricular collagen deposition ranged from 0.4 (7 mo) to 3.5 (32-48 mos) (Fig. 1).

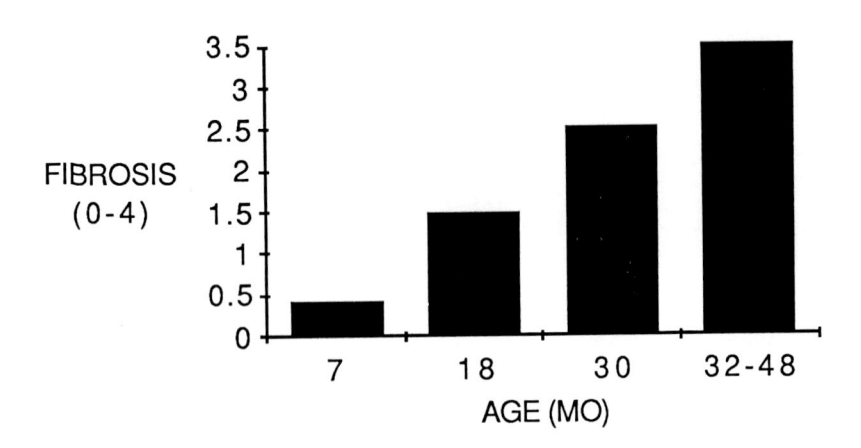

Figure 1. The degree of fibrosis (0-4) for rats in the CV-F group.

The fibrosis was most dominant in the left ventricle, although the right ventricle and both atria were involved to a more limited degree. The collagen was present as focally diffuse deposits, predominantly in the ventricular subendothelium. The Masson staining was dense and intensely blue, especially in the older rats, indicating a high degree of cross linking. Blood vessels had normal wall thickness without endothelial plaques or narrowed lumena. Heart valves were free of scarring. No increase in collagen was present in lung, liver or kidney tissues.

The hearts of CV-R rats contained less collagen than CV-F rats for each age group (Fig. 2a, 2b).

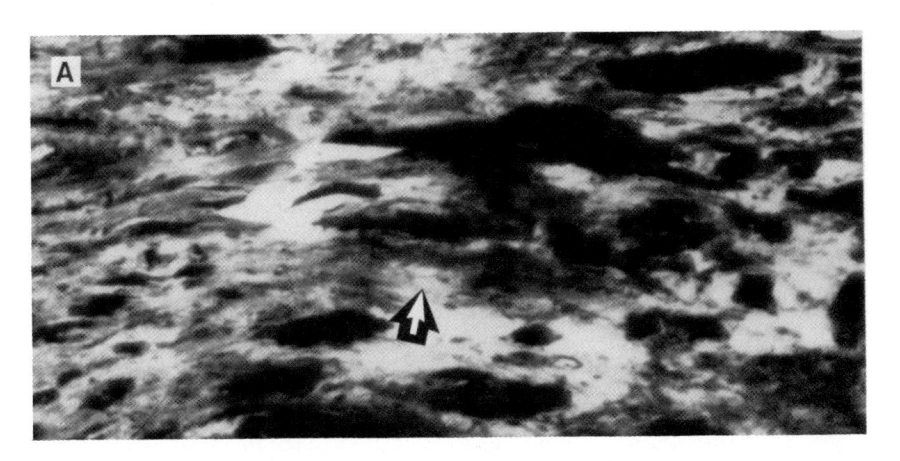

Figure 2: Legend on the following page.

Figure 2. Photomicrographs of 6u paraffin sections of rat ventricles treated with Masson trichrome stain (x300).
a. conventional fullfed (CV-F) at 34 months; b. conventional restricted diet (CV-R) at 33 months. Arrows point to collagen surrounding or infiltrating myofibrils.

This difference was statistically different when all diet restricted (CV-R + CR-R) were compared with all fullfed (CV-F + GF-F) rats in the middle (18 mo:P=.02) and late age (30 mo:P=.002; 32-48 mo:P=.006) groups (Fig. 3). A similar comparison of all germfree vs. conventional rats showed no difference at any age level (P $\geq$ 0.1).

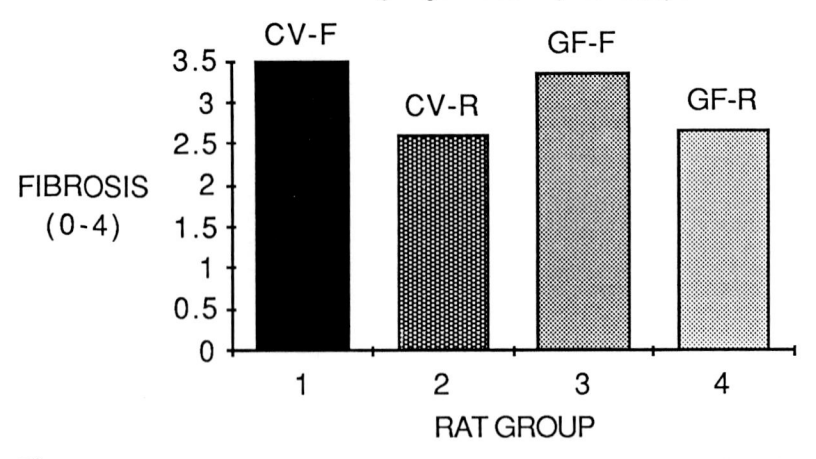

Figure 3. The average degree of cardiac fibrosis (0-4) of each group of rats in the 32-48 mo age range (CV-F: conventional fullfed; CV-R: conventional restricted diet; GF-F: germfree fullfed; GF-R: germfree restricted diet).

DISCUSSION

Unlike the human heart (Oken and Boucek, 1957), the aging rat heart contains significant deposits of collagen (Mohan and Radha, 1980; Cappelli et al., 1984). The cause of cardiac fibrosis is unknown, although there is some evidence that collagen-degrading activity of rat heart muscle decreases during the aging process. There is no reported evidence that the cardiomyopathy is caused by coronary vessel disease (Oken and Boucek, 1957).

The influences of germfree environment and diet restriction on cardiomyopathy in the aging rat have not been studied extensively. Diet restriction has been shown to reduce cardiomyopathy in the Fischer 344 rat (Maeda et al., 1985). However, heart disease in this strain of rat has been associated with chronic nephropathy, a disorder which is also reduced by restricted diet. In view of recent studies that soy-based diet reduces the severity of nephropathy in Fischer rats (Iwaski et al., 1988), further studies of diet restriction in that strain are required. The present study confirms the age-related collagen deposit in rat heart and provides clear evidence in the Lobund-Wistar rat that restriction of diet partially protects against this phenomenon.

ACKNOWLEDGEMENT

Technical assistance of Beverly Thomas and preparation of the manuscript by Lynn Gibson are appreciated. This study was supported in part by the W.P. Cornwell Amyloid Research Fund.

Cappelli V, Forni R, Poggesi C, Reggiani C, Ricciardi L (1984). Age-dependent variations of diastolic stiffness and collagen content in rat ventricular myocardium. Archives Internationales de Physiologic et de Biochimie 92:93-106.
Cochran WE, Snedecor GW (1980). Statistical Methods. 7th edition. Ames, Iowa: The Iowa State University press, pp 215-237.
Gomori G (1937). Silver impregnation of reticulum in paraffin sections. Amer J Path 13:993-1002.
Iwaskai K, Gleiser CA, Masoro EJ, McMahan CA, Seo E, Yu BP (1988). The influence of dietary protein source on longeivty and age-related disease processes of Fischer rats. J Gerontol:Biol Sci 43:B5-B12.
Lillie RD (1985). Histopathologic Technic and Practical Histochemistry. 3rd edition. New York: McGraw-Hill Book Co., p 547.
Maeda H, Gleiser CA, Masoro EJ, Murata I, McMahan CA, Yu BP (1985). Nutritional influences on aging of Fischer 344 rats. II. Pathology. J Gerontol 40:671-688.

Masson PJ (1929). Trichrome stainings and their preliminary technique. J Techn Methods 12:75-90.

Mohan S, Radha E (1980). Age-related changes in rat muscle collagen. Gerontology 26:61-67.

Oken DE, Boucek RJ (1957). Quantitation of collagen in human myocardium. Circulation Research, 5:357-361.

Puchtler H, Sweat F (1965). Congo red as a stain for fluorescence microscopy of amyloid. J Histochem Cytochem 13:693-694.

Sainte-Marie G (1962). A paraffin embedding technique for studies employing immunofluorescence. J Histochem Cytochem 10:250-256.

Shirahama T, Skinner M, Cohen AS (1981). Immunocytochemical identification of amyloid in formalin-fixed paraffin sections. Histochemistry 72:161-171.

Sternberger LA (1979). Immunocytochemistry. 2nd edition. New York: John Wiley and Sons.

Westermark P, Johansson B, Natvig JB (1979). Senile cardiac amyloidosis: the existence of two different amyloid substances in the aging heart. Scand J Immunol 10:303-308.

Westermark P, Natvig JB, Johansson B (1977). Characterization of an amyloid fibril protein from senile cardiac amyloid. J Exp Med 146:631-636.

Winer BJ (1971). Statistical Principles in Experimental Design, 2nd edition. New York: McGraw-Hill Book Co., p 201-204.

EFFECTS OF AGING, DIET RESTRICTION AND MICROFLORA
ON ORAL HEALTH IN HUMANS AND ANIMALS

Sam Rosen, Mike Strayer, William Glocker, James Marquard and F. Mike Beck

College of Dentistry, The Ohio State University, Columbus, Ohio 43210

INTRODUCTION

Normal aging is an inevitable process resulting in a finite life span for each living organism. The loss of physiological function associated with aging is linear. Consequently an eighty year old ages at the same rate as a thirty year old. However the older individual appears more aged due to the accumulation of age related changes which are secondary to the aging process. A great variability or diversity of change is the single most important characteristic of aging. This variability of change is found within individuals as well as between individuals (Gilchrest and Rowe, 1982).

The study of aging has prompted wide ranging research into the physiological, psychological, and sociological aspects of normal aging. Aging is often associated with decrements in performance, but impairment of functions is not a normal consequence of aging. Although everyone ages, disease occurs only in a portion of the aged population. While the incidence of disease increases with age, aging and disease are not synonymous (Shock, 1984).

Limited data are available regarding the relationship between aging and oral physiology. Common generalizations reported in the literature on the aging oral cavity include: the inevitable loss of dentition; atrophy of the oral mucosa; increasing prevalence of cervical caries; atrophy of the oral-facial musculature; alterations in the amount and composition of saliva; the wearing of the hard tissue; and altered sensory function (taste, smell, touch). However the literature which reports these generalizations was found to have many methodological and research design faults (Baum, 1981a).

The NIA's Baltimore Longitudinal Study of Aging has examined taste acuity, taste perception, oral motor function, and salivary gland function among its participants to judge the validity of early generalizations regarding the aging oral cavity. Over 70% of the participants in the Baltimore Longitudinal Study of Aging take no prescription medication. This is an indication of the low incidence of systemic disease among the study subjects (Baum, 1981b). Among the non-medicated males and females no diminution in parotid fluid output with age was found. Preliminary findings indicated that 1 in 5 individuals, 60 or older, showed some decrease in oral muscular function which could affect competent swallow. An increase in the prevalence of lip droop associated with age was also reported. This appears to be a common problem among individuals over 80 years. Additional findings indicated there were specific altered taste performances in older individuals. It was suggested that post-menopausal females require higher concentrations of salt and sucrose to detect their presence in food (Baum and Bodner, 1983).

The impact of aging on the oral mucosa is not well understood. A review of the literature found conflicting data concerning the microscopic features and the kinetic and metabolic activities of the skin and oral mucosa in aged animals (Hill, 1984). A variety of changes in the oral mucosa are associated with nutritional deficiencies. Poor nutrition can manifest itself through oral changes associated with systemic conditions or through oral disease, such as coronal carries, root caries or periodontal disease. The polypharmacy frequently found among the elderly further complicates their oral health status. This is due to the xerostomic side effects associated with many prescription medications (Papas, 1984).

Dental decay is another issue of great concern for older adults. Caries rates in adults are different than that found in children and adolescents. In adults, secondary (recurrent) caries is proportionately higher than primary caries. There is also a greater prevalence of root caries in older adults. This is due to a combination of factors which leads to the apical migration of the periodontal attachment (Beck, 1984; Banting, 1984). While root caries has been identified as a growing concern for older adults, it is not a recent phenomenon. Root caries was the predominant type of caries present prior to the seventeenth century (Beck, 1984). Data from the recent NIDR Adult Oral Health Survey showed increasing mean DFS (root surface) rates with age for participants over the age of 65 (1987). Hand and Hunt reported that 44% of older adults developed new root caries averaging of 1.1 surfaces per person during a 36-month caries incidence study (Hand and Hunt, 1988).

Trends in periodontal disease, drawn from population-based epidemiologic studies, have shown no significant increase in mean periodontal index scores over time for 65-74 year olds. The proportion of 65-74 year olds without periodontal disease increased approximately 4 fold for males and 3 fold for females over a 12 year period. An increase in the proportion of older adults free of periodontal disease was reported to be a result of a shift from category of periodontal disease without pockets to the category of absence of periodontal disease (Katz and Meskin, 1986). With more teeth at risk for developing periodontal disease, it has been suggested but not well documented that there is an increased susceptibility to periodontal disease with age (Page, 1985). Beck, in a review of the dental epidemiology literature, notes that at best the epidemiology of periodontal disease in all age groups is confused (Beck, 1984).

With the difficulty of finding a healthy, non-medicated older adult population, the use of an animal model to study the effects of aging on the oral cavity is highly desirable. The lack of animal susceptibility to periodontal disease showing a range of symptoms similar to humans has hampered research efforts in this field (Thilander, 1961; Jordon, 1971). Currently various rodent species have proven to be adequate, if not ideal, models for periodontal research (Baer and Fitzgerald, 1966; Belting et al, 1953; Gilmore and Hickman, 1959; Gupta and Shaw, 1956a; Irving et al, 1974).

Advancing age, diet and microflora are linked in the pathogenesis of periodontal disease. Several methods have been used to evaluate and describe its pathologic features. Bone loss estimates have been expressed as the distance between the dental cemento-enamel junction and alveolar bone crest (Costich, 1955; Guggenheim and Schroeder, 1974; Gupta and Shaw, 1956b; Keyes and Gold, 1955). In addition, histopathology (Mulvihill et al., 1967), histochemistry, electron microscopy (Garant, 1976), autoradiography (Irving et al, 1975), and immunology (Guggenheim and Schroeder, 1974) studies have been used in the evaluation of periodontal disease.

Studies evaluating microflora and age effects on periodontal disease have been accomplished using both conventional and germ-free animals (Jordon et al., 1972; Amstad-Jossi and Schroeder, 1978). A recent report by Pollard and Wostmann (1985) shows that a germ- free state produces age longevity over conventional animals. With 25% restriction of food intake, germ-free animals show increased longevity and an apparent freedom of systemic diseases.

Both conventional and germ-free animals on _ad lib_ diets are susceptible to and usually succumb to various neoplastic and renal diseases. The objective of this study is to evaluate the oral health status of conventional and germ-free Lobund-Wistar (L-W) rats of varying ages which are divided into groups receiving _ad lib_ and restricted diets (25% dietary restriction).

METHODS

Heads of Lobund-Wistar (L-W) rats which were sacrificed or died naturally were supplied by the Lobund Laboratory, University of Notre Dame. Only males were used in this study. The animals were divided into four groups: conventional full-fed (CF); conventional restricted (CR); germ-free full-fed (GF); and germ-free restricted (GR).

As the heads were received, the upper and lower jaws were divided into a total of 4 segments by splitting each jaw through mid suture lines. One mandibular segment was used for determination of alveolar bone loss (Doff et al, 1977). The jaws were defleshed and stained with murexide (ammonium purpurate) to determine decalcified areas.

Oral health factors studied were: coronal caries, root surface caries, alveolar bone loss, occlusal wear, and bone density. In addition, salivary gland weights were determined. The submandibular glands were surgically recovered, hemisected, and processed for routine paraffin sectioning. Step serial sections were cut at 5 microns and mounted on glass slides using gelatin as the mounting adhesive. Selected slides were stained with: 1) Mayer's hematoxilyn and eosin, 2) Southgare's modification of Mayer's mucicarmine, 3) periodic acid-Schiff with hematoxilyn counterstain, and 4) phloxine-tartrazine for cytoplasmic granules. All observations were made with light microscopy. A total of 39 heads have been examined at age levels of 3, 18 and 36 months.

RESULTS

Neither coronal nor root surface caries were detected in any of the rats. Occlusal wear and alveolar bone loss were clearly evident in the 18 and 36 month rats. Mean alveolar bone loss scores of mandibular molars are given in Table 1. Mean weights of submandibular salivary glands of 3 month old rats are given in Table 2.

Histological comparisons of the submandibular glands of 3 month and 36 month old rats revealed no qualitative differences in mucous acini, excretory ducts, serous cells, striated ducts, and intercalated ducts. However, a marked difference was observed in the numbers of granular ducts (Fig. 1) in the 36 month group. This change is consistent with a replacement of acini by granular ducts in the older animals. The lack of significant changes in the gross weight of the glands of the two groups supports this observation.

From Table 1, it may be seen that age had a significant effect on alveolar bone loss scores ($F=85.1$, $p=0.0001$); but that group did not ($F=0.83$, $p=0.487$). In addition, there was no significant age x group interaction ($F=0.84$, $p=0.533$).

Table 1. Mean alveolar bone loss scores in Lobund-Wistar rats.

Age in mos.	CF	CR	GF	GR	Total*
3	22.0±0.00 (2)	23.0±2.00 (2)	23.0±3.00 (2)	23.0±3.00 (2)	22.8±0.90 (8)
18	34.7±2.19 (3)	30.3±2.96 (3)	41.0±4.14 (4)	36.3±1.80 (4)	36.0±1.70 (14)
36	52.3±4.13 (4)	52.8±1.44 (4)	52.0±1.53 (3)	50.5±2.38 (6)	52.7±1.24 (17)
TOTAL+	39.7±4.66 (9)	38.7±4.68 (9)	40.7±4.16 (9)	41.2±3.38 (12)	

Age, $F=85.1$, $P=.0001$
Group, $F=0.83$, $P=0.487$
Age x Group, $F=.84$, $P=0.553$
*Newman-Keuls shows all means significantly different ($p<0.05$)
+Newman-Keuls shows no significant difference ($p>0.05$)
( ) = sample size

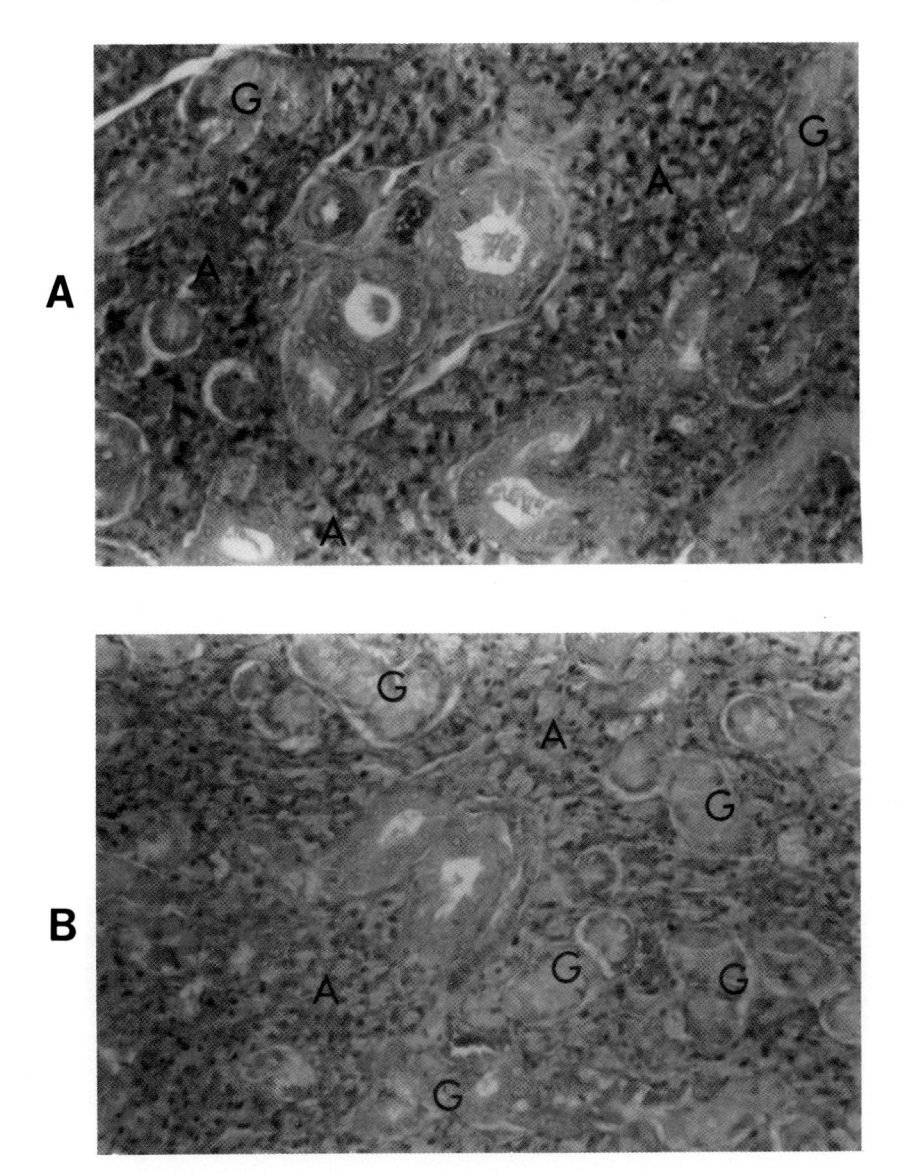

Figure 1. A is representative of a submandibular gland from a three month old animal. Note the density of glandular acini (a) (original mag, 250X). B is representative of a 36 month old animal. There appears to be fewer acini per unit of area with apparent replacement by granular duct (g) proliferation. (original mag, 100X).

Table 2. Mean submandibular gland weight (mg) ± S.E. at 3 months of age.

CR	GR	CF	GF
<u>464 ± 22.4</u>	<u>470 ± 7.50</u>	573 ± 23.6	810 ± 20.5
(6)	(2)	(4)	(5)

Underlined means are not significantly different (p>0.05)
( ) = sample size

Bone density was qualitatively observed and it was found that rats 3 months old showed less density than the older animals indicating that at this age the jaws were not fully calcified (Fig. 2).

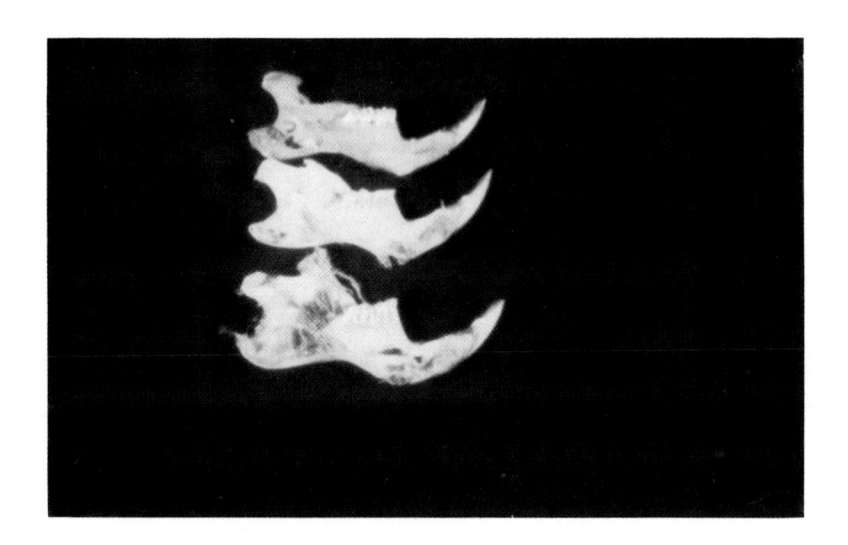

Figure 2. Bone loss and bone density in mandibles of rats. Top = 3 months old; Middle = 18 months old; and Bottom = 36 months old. The dark stain (murexide) in the 3 months old mandible indicates less calcification.

DISCUSSION

Bone loss scores were associated with age more so than with any factor. In 18 month old animals less bone loss was observed in animals on a restricted diet, but this decrease was not significant. Hair impactions in the gingival sulcus were not evident. Since bone loss occurred in rats with and without microorganisms present, it is suggested that the fibrous nature of the diet was a contributor to this condition. However, the prime contributor to alveolar bone loss is likely an unknown aging factor.

Data for salivary gland weight are given only for 3 month old rats, since sufficient numbers were not available for the other age groups. The significant increase in salivary gland weight for full fed animals corresponds to their greater body weights (Snyder D, University of Notre Dame, personal communications, 1987). However, we are unable to explain why there is a significant increase in salivary gland weight in germ-free full-fed over conventional full-fed animals since their body weights were not significantly different.

REFERENCES

Amstad-Jossi M, Schroeder HE (1978). Age-related alterations of periodontal structures around the cemento-enamel junction and of the gingival connective tissue composition in germ-free rats. J Periodontal Res 13:76-90.
Baer PN, Fitzgerald RJ (1966). Periodontal disease in the 18-month-old germfree rat. J. Dent Res 45:406.
Banting DW (1984). Dental caries in the elderly. Gerodontology 3:55-61.
Baum BJ (1981a). Current research on aging and oral health. Spec Care Dent 1:105-109.
Baum BJ (1981b). Characteristics of participants in the oral physiology component of the Baltimore Longitudinal Study of Aging. Comm Dent Oral Epidemiol 9:128-134.

Baum BJ, Bodner L (1983). Aging and oral motor function: Evidence for altered performance among older adults. J Dent Res 62:2-6.

Beck JD (1984). The epidemiology of dental disease in the elderly. Gerodontology 3:5-15.

Belting CM, Schour L, Weinmann JP, Shepro MV (1953). Age changes in the periodontal tissues of the rat molar. J Dent Res 32:332-353.

Costich ER (1955). A quantitative evaluation of the effect of copper on alveolar bone loss in the Syrian hamster. J Periodontal Res 26:301-305.

Doff RS, Rosen S, App G (1977). Root surface caries in the molar teeth of rice rats. I. A method for quantitative scoring. J Dent Res 56:1013-1016.

Garant PR (1976). An electron microscopic study of the periodontal tissues of germfree rats and rats monoinfected with Actinomyces naeslundii. J Periodontal Res 11, Suppl1 No. 15, pp 9-79.

Gilchrest B, Rowe J (1982). The biology of aging. In Rowe J and Besdine R (eds.): "Health and disease in old age." Boston:Little, Brown and Company, pp 15-19.

Gilmore ND, Hickman I (1959). Some age changes in the periodontium of the albino mouse. J Dent Res 38:1195-1206.

Guggenheim B, Schroeder HE (1974). Reactions in the periodontium to continuous antigenic stimulation in sensitized gnotobiotic rats. Infection and Immunity 10:565-577.

Gupta OP, Shaw JH (1956a). Periodontal disease in the rice rat. I. Anatomic and histopathologic findings. Oral Surg 9:595-603.

Gupta OP, Shaw JH (1956b). Periodontal disease in the rice rat. II. Methods for the evaluation of the extent of periodontal disease. Oral Surg 9:727-735.

Hand JS, Hunt RJ (1988). Coronal and root caries in older adults: 36 month incidence. J Dent Res Vol 67 (Special Issue):176, Abstract No. 507.

Hill M (1984). The influence of aging on skin and oral mucosa. Gerodontology 3:35-45.

Irving JT, Socransky SS, Heeley JD (1974). Histologic changes in experimental periodontal dis-

ease in gnotobiotic rats and conventional hamsters. J Periodontal Res 9:73-80.
Irving JT, Newman MG, Socransky SS, Heeley JD (1975). Histological changes in experimental periodontal disease in rats mono-infected with a gram-negative organism. Arch Oral Biol 20:219-220.
Jordan HV (1971). Rodent model systems in periodontal disease research. J Dent Res 50:236-245.
Jordan HV, Keyes PH, Bellack S (1972). Periodontal lesions in hamsters and gnotobiotic rats infected with actinomyces of human origin. J Periodontal Res 7:21-28.
Katz RV, Meskin LH (1986). The epidemiology of oral diseases in older adults. In Holm-Pedersen P and Loe H (eds.): "Geriatric dentistry." Copenhagen:Munksgaard, p 222.
Keyes PH, Gold HS (1955). Periodontal lesion in the Syrian hamster. I. A method of evaluating alveolar bone resorption. Oral Surg 8:492-499.
Mulvihill JE, Susi FR, Shaw JH, Holdhaber P (1967). Histological studies of the periodontal syndrome in rice rats and the effects of penicillin. Arch Oral Biol 12:733-744.
National Institute of Dental Research (1987). Oral health of United States adults, national findings. NIH Publication No 87-2868:111.
Page RC (1985). Oral health status in the United States: Prevalence of inflammatory periodontal disease. J Dent Ed 49:354-364.
Papas A (1984). Oral health status of the elderly, with dietary and nutritional considerations. Gerodontology 3:147-155.
Pollard M, Wostmann BS (1985). Aging in germfree rats: The relationship to the environment, disease of endogenous origin, and to dietary modification. In Archibald J, Ditchfield J, Rowsell HC (eds): "The Contribution of Laboratory Animal Science to the Welfare of Man and Animals," New York: Gustav Fischer Verlag, pp 181-186.
Shock N (ed.) (1984). Normal human aging: the Baltimore Longitudinal Study of Aging. Dept of Health and Human Services, NIH Publication No 84-2450:1.
Thilander H (1961). Periodontal disease in the white rat. Trans R Sch Dent Stockh 6:1-99.

AGE-RELATED CHANGES IN ORGANIC AND INORGANIC BONE MATRIX CONSTITUENTS: THE EFFECT OF ENVIRONMENT AND DIET

Satoru K. Nishimoto and Steven M. Padilla

Laboratory of Connective Tissue Biochemistry, Orthopaedic Hospital, Department of Orthopaedics, University of Southern California, Los Angeles, California 90007

INTRODUCTION

The age-related decline in bone mass is a well established phenomenon. An acceleration of this process can lead to osteoporosis, a clinically important disease. Osteoporosis is common in elderly women, and results in significant morbidity and mortality. The causes of osteoporosis remain incompletely understood, but are clearly a result of an imbalance between bone resorption and bone formation.

We have previously shown that the rate of bone formation decreases with increasing age in the rat (Nishimoto et al., 1985). In the same study, we showed that the content of the bone-carboxyglutamic acid protein (bone Gla protein, BGP) in bone also declines between 1, 3 and 9 months of age (Nishimoto et al., 1985). We have also demonstrated that serum BGP levels also declined with age (Nishimoto et al., 1985). The serum BGP has been proposed as a reflection of the underlying state of bone metabolism (Price and Nishimoto, 1980; Price et al., 1980). Studies by others have demonstrated that magnesium levels in bone decline with age, as does calcium, the most abundant mineral (McDonald et al., 1986; Kiebzak et al., 1988). BGP levels in bone matrix decline with age, while collagen, the major bone matrix constituent, does not change significantly with age (Nishimoto et al., 1985; McDonald et al., 1986; Kiebzak et al., 1988). Mechanically, the maximum breaking force required to fracture femurs at midshaft does not change with age, while ultimate stress, a parameter which normalizes bone for differences in size and geometry, decreased (Kiebzak et al., 1988). Thus, age-related changes in bone matrix appear to be dependent upon the parameter measured.

Food restriction of rats is known to modulate many, but not all, age-associated processes. Using the germfree and conventionally raised Lobund-Wistar (L-W) rat strain on restricted and free-fed diets, we tested for effects of a germfree environment or of dietary restriction on four parameters of the bone matrix. We measured calcium and magnesium as major inorganic bone matrix components, and collagen and bone Gla as major organic components.

MATERIAL AND METHODS

Tibia samples from rats were received frozen. At the time of sacrifice, bone ends were removed and the marrow removed by flushing with distilled water. The midshafts received were cleaned of adhering connective tissue, rinsed with distilled water, and freeze-dried. The bones were then ground in a Spex freezer mill.

Bone Gla protein determination. The dry bone powder was mixed with 10% formic acid (10 ul/mg) and extracted for 20 h at 4 C. 10 ul of the supernatant was removed after centrifugation, neutralized with 50 ul of 1% NaOH in 0.5 M Na_2HPO_4, and diluted to a final 1:1060 with RIA diluent (0.01 M Na_2HPO_4, 0.14 NaCl, 0.1% Tween 20). The BGP assays were performed on this dilution as previously described (Nishimoto et al., 1985). The remainder of the formic acid extract (containing the insoluble bone residue) was transferred quantitatively to an ashing crucible and freeze-dried. These were used for mineral analysis as described below.

Calcium and Magnesium Determination. The freeze-dried formic acid extract and residues were ashed at 700 C overnight. The calcium and magnesium were determined from the 6 N HCl resolubilized ash after appropriate dilution in lanthanum oxide solution as described (Nishimoto et al., 1985). Calcium was determined by atomic absorption spectroscopy at 422.7 nm, and magnesium at 285.2 nm on a Perkin-Elmer model 560 atomic absorption spectrophotometer.

Hydroxyproline determination. Hydroxyproline was determined in the formic acid insoluble residue by a colorimetric assay (Cheung and Nimni, 1984). Briefly, 40-50 mg of dry bone was extracted for 20 h with 10 ul/mg of 10% formic acid at 4 C. The insoluble residue was then washed twice and freeze-dried. The freeze-dried residue containing the majority of bone collagen was then hydrolyzed to free amino acids with 6 N HCl. The hydrolyzates were dried, and dilutions of the resolubilized hydrolyzate were assayed as described (Cheung and Nimni, 1984).

Determination of serum BGP. Circulating levels of bone Gla protein were determined by RIA as previously described (Nishimoto et al., 1985). Briefly, 25 ul of serum removed at the time of sacrifice was assayed in triplicate.

Statistical analysis. The effect of age, diet, or germfree environment was analyzed using single factor analysis of variance (general linear models procedure, SAS Institute, 1987 edition, on a DEC VAX 11/750). Differences between groups were determined using Duncan's multiple range test.

RESULTS

Frozen tibia samples were received from the Lobund Laboratory after scheduled sacrifice at 6,

18, and 30 months. Four groups of rats were maintained for this period: a conventionally raised, free-fed group (CF); a conventionally raised, restricted diet group (CR); a germfree, free-fed group (GF); and a germfree, restricted diet group (GR) (see the chapter by Snyder and Wostmann in these proceedings).

The effects of dietary restriction and/or germfree environment on the concentration of the bone specific protein BGP are shown in figure 1. A general decline in BGP level is seen from 6 months through 30 months. When rats are grouped by age alone, analysis of variance and the Duncan test show a significant difference between 18 month and 30 month old pooled groups. This result correlates well with previously published data (Nishimoto et al., 1985; Kiebzak et al., 1988). Analysis of variance also demonstrated no significant differences between the 4 treatment groups, with two exceptions. The CR group at 18 months was significantly different from the others, and the GF group at 30 months was found to be significantly different ($p<0.05$).

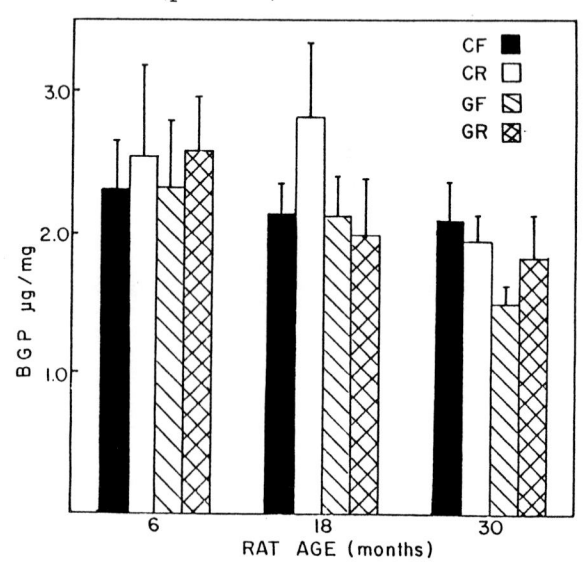

Figure 1. Bone Gla protein content in tibial midshafts of L-W rats.

Magnesium values for all four groups of rats show a consistent, significant decline with age (p<0.05, figure 2). Analysis of variance showed no differences in bone magnesium content between treatment groups at any age. This result correlated well with previous reports of a decline in bone magnesium content with age (McDonald et al., 1986).

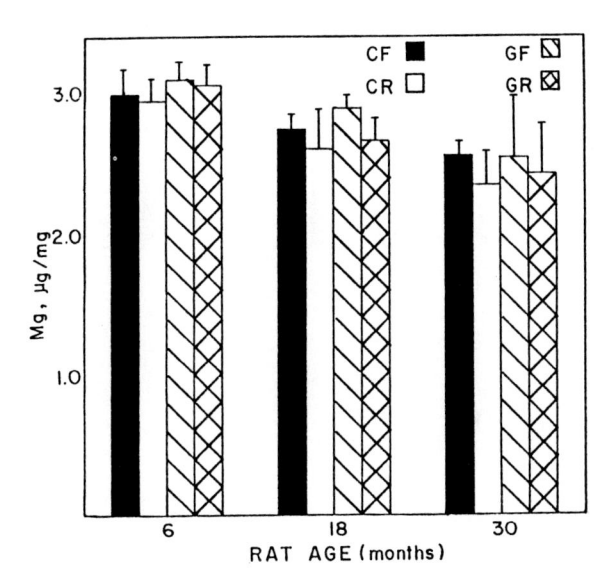

Figure 2. Magnesium content in tibial midshafts of L-W rats.

Calcium values declined significantly (p<0.05, figure 3) between 6 and 18 months, but were not found to change between 18 and 30 months. No differences were noted between the 4 treatment groups of experimental rats.

Bone collagen was assessed by measurement of bone hydroxyproline, an abundant post-translationally modified amino acid in collagen. Hydroxyproline levels did not change significantly with age, and no differences were observed between experimental groups (figure 4).

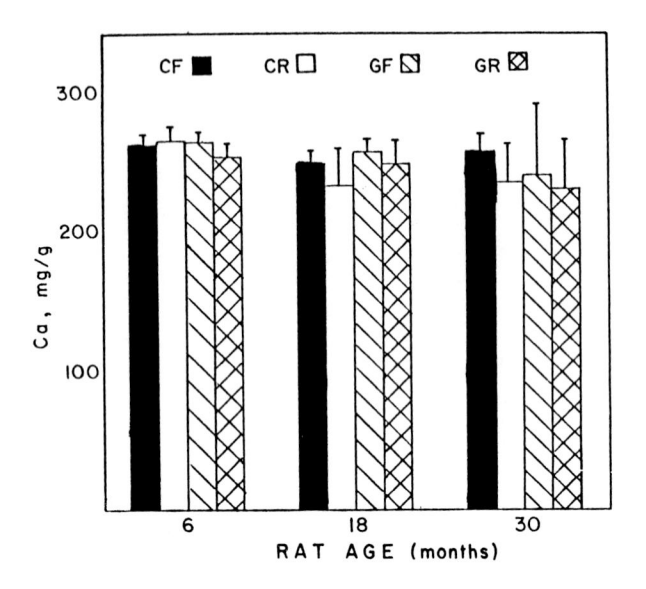

Figure 3. Calcium content in tibial midshafts of L-W rats.

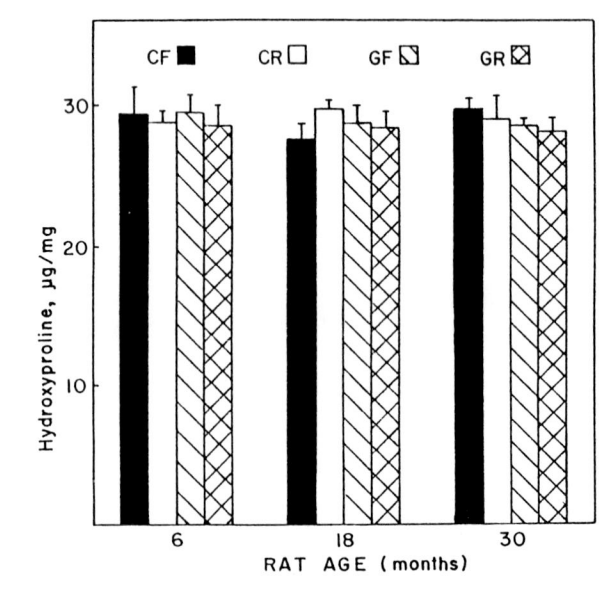

Figure 4. Hydroxyproline content in tibial midshafts of L-W rats.

Finally, we determined the serum level of circulating BGP in these rats, and found an age-related decline in the serum BGP concentration (figure 5). This correlated well with our previous findings, and also with another recently published study (Nishimoto et al., 1985; Kiebzak et al., 1988).

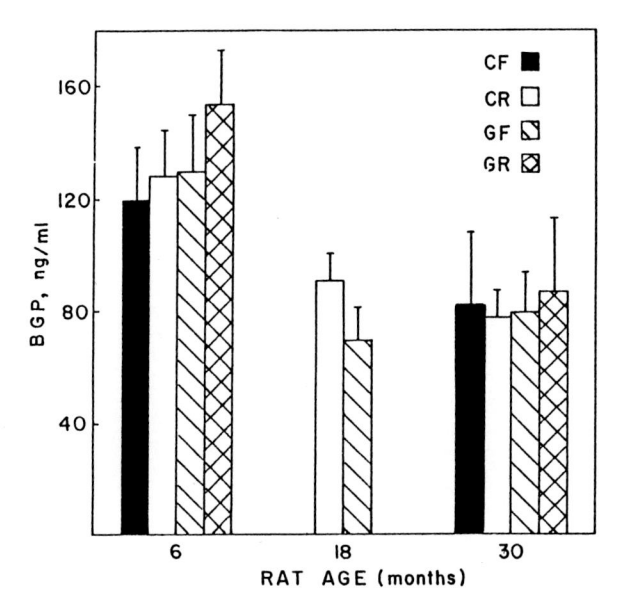

Figure 5. Serum bone Gla protein levels in L-W rats.

DISCUSSION

Bone undergoes complex changes with age. Several parameters and constituents of the bone matrix appear unaffected, while others show an age-related decline (Nishimoto et al., 1985; McDonald et al., 1986; Kiebzak et al., 1988). We have studied several bone matrix markers including serum bone Gla protein in an attempt to assess the effect of a germfree environment or dietary restriction on these markers, so that we might infer the ability of these treatments to increase the capacity of the senescent animal to form bone or to repair it. Our results imply that neither a germfree environment nor dietary restriction reverse or even slow the changes which occur in bone. Further studies to directly test the bone forming capacity of the restricted diet rat compared to its full-fed counterpart might be accomplished by measurement of the bone formation rate after implantation of bone inductive materials as we have previously described (Nishimoto et al., 1985).

REFERENCES

Cheung DT, Nimni ME (1984). Mechanism of crosslinking of proteins by glutaraldehyde III. Reaction with collagen in tissue. Connect Tissue Res 13:109-115.
Kiebzak GM, Smith R, Gundberg CC, Howe JC, Sacktor B (1988). Bone status of senescent male rats: Chemical, morphometric, and mechanical analysis. J Bone and Mineral Res 3:37-44.
McDonald R, Hegenauer J, Saltman P (1986). Age-related differences in the bone mineralization pattern of rats following exercise. J Gerontol 41:445-452.
Nishimoto SK, Chang C-H, Gendler E, Stryker WF, Nimni ME (1985). The effect of aging on bone formation in rats: Biochemical and histological evidence for decreased bone formation capacity. Calcif Tissue Int 37:617-624.
Price PA, Nishimoto SK (1980). Radioimmunoassay for the vitamin K-dependent protein of bone and its discovery in plasma. Proc Natl Acad Sci 77:2234-2238.
Price PA, Parthemore JG, Deftos LJ (1980). New biochemical marker for bone metabolism. J Clin Invest 66:878-883.

# IMMUNOLOGY

HOW DOES DIETARY RESTRICTION RETARD DISEASES AND AGING?

Richard Weindruch

Biomedical Research and Clinical Medicine Program, National Institute on Aging, NIH, Bethesda, MD 20892

INTRODUCTION

Dietary restriction (DR) of caloric intake but without essential nutrient malnutrition is a quite simple experimental maneuver which produces many desirable biological outcomes (see Holehan & Merry, 1986; Masoro, 1988 & this volume; Weindruch & Walford, 1988). DR extends average and maximum life span in diverse species and, in rats and mice, delays disease onset and slows the rates of change for most of the age-sensitive biologic parameters tested to date. These effects of DR on rodents are unmatched by competitor methods. Thus, DR is the best model available to study the biology of decelerated aging.

As a result of DR's actions, the title of this brief article asks one of the two main questions challenging the growing DR field (The other question is: Does DR slow the rate of aging in primates?). The present comments briefly summarize some of the views co-generated with Roy Walford during our 12 year collaboration. For a more extensive consideration, see Weindruch & Walford (1988).

MECHANISTIC POSSIBILITIES FOR DR'S ACTIONS

Two main facts need to be kept in mind when considering DR's mode(s) of action. One is the apparent phylogenetic independence of the effect. It is not simply a "rodent phenomenon"! Dietary restriction extends life span in nearly all species so far tested (the few exceptions seem to be special cases), from primitive animals such as protozoans to complex ones such as rodents. A second main fact is that only energy restriction retards aging in mammals. Without calorie restriction, maximum life span has not been prolonged beyond that which is usual for the species by other dietary adjustments limited to protein, fat, carbohydrate, vitamin or mineral content.

It is also important to consider the existence of age-sensitive but DR-resistant parameters (Walford et al., 1987). Nearly all of the age-sensitive biological parameters studied to date in DR rodents stay "younger longer." The few DR-resistant ones may have little to do with basic aging (even if they change with age) or they may be responsible for DR animals not living even longer. Thus, in the DR paradigm, so-called "negative data" may be important data. Table 1 lists age-sensitive but apparently DR-resistant parameters.

TABLE 1. AGE-SENSITIVE BUT DR-RESISTANT PARAMETERS*

Parameter	Age Change	Ref.**
Circulating TSH (thyrotropin)	↓	1
Insulin secretion (islets in vitro)	↓	2
Insulin secretion (per vol of islet)	↓	3
Antilipolytic effect of insulin	↓	4
Oxidative damage products in urine	↑	5
Blood pressure	↑	6
Forskolin-stim. adenylate cyclase activ.	↓	7
Protein synthetic activity in brain	↓	8
β-oxidation in liver	↓	9

*Certain of these findings require repetition using DR regimens proven to extend maximum life span.
**Complete references are in Weindruch & Walford (1988). (1) Sarkar et al. (1982); (2) Reaven & Reaven (1981b); (3) Reaven et al. (1983a); (4) DiGirolamo et al. (1984); (5) Saul et al. (1987); (6) Yu et al. (1985); (7) Scarpace & Yu (1987); (8) Sparks et al. (1983); (9) Rumsey et al. (1987).

Future work aimed at elucidating mechanisms behind DR's actions must not just show by ever more sophisticated means that DR animals are functionally younger. For example, there is a tendency to reason that because protein synthesis is sometimes greater (=functionally "younger") in DR animals, the mechanism of DR's action is to increase protein synthesis. But such an increase in protein synthesis may merely mean that the animal has been kept functionally "younger" by some other non-protein synthesis mechanism.

To elucidate how DR retards aging and extends life span is no easy task, especially in view of the present level of understanding on the etiology of aging. However, with this caveat in mind, my comments on certain of the most likely possibilities follow.

Energy Metabolism

All DR regimens which extend maximum life span limit energy intake. Thus, major energy-producing and -consuming processes, as well as processes which detoxify noxious byproducts of energy metabolism, merit serious attention in considering DR's basic mechanisms.

Four non-mutually exclusive, energy metabolism-linked possibilities are that DR acts by: i) lowering the metabolic rate at some critical but unappreciated level (e.g. per whole animal or per total mass of internal organs) or locale (e.g per brain or per physiologic "sensing center"); ii) increasing metabolic efficiency; iii) decreasing heat and free radical production, with less damage to mitochondria and/or other sites; and iv) increasing detoxification of metabolically-generated active oxygen.

The finding by McCarter et al. (1985) that O_2 consumption per lean body weight of rat is unaltered by long-term DR seems to be inadequate grounds on which to abandon metabolic rate reduction as contributing to DR's effects. How closely does O_2 consumption by the intact animal estimate actual rates of oxidative metabolism? What per cent of the total O_2 consumed is used for mitochondrial energy metabolism versus O_2 used otherwise (e.g. mixed function oxidases, oxygenases)? Clearly, O_2 consumption is much lower (~30%) per whole DR animal. Also, O_2 use needs to be considered on a tissue-by-tissue basis because DR strongly reduces the weight of most tissues (e.g. lymphoid tissues, muscle, fat) but not others (e.g. brain).

Metabolic efficiency can be viewed from two standpoints. The efficiency of energy generation represents the fraction of the energy ingested which is absorbed and trapped in a biologically useful form (Fig. 1). The <u>efficiency of energy usage</u> can be thought of as the percentage of trapped energy used for essential physiologic functions, versus that wasted as heat generation, or in maintaining the tissues of an obese animal, or in other non-essential ways. Perhaps, much of DR's effect on aging is due to an increase in the efficiency of energy generation and/or use. Investigation of this possibility by researchers of energy metabolism and metabolic efficiency (e.g. the basis of energy coupling during ATP formation, cellular energy costs, thermogenesis, futile cycles) appears timely.

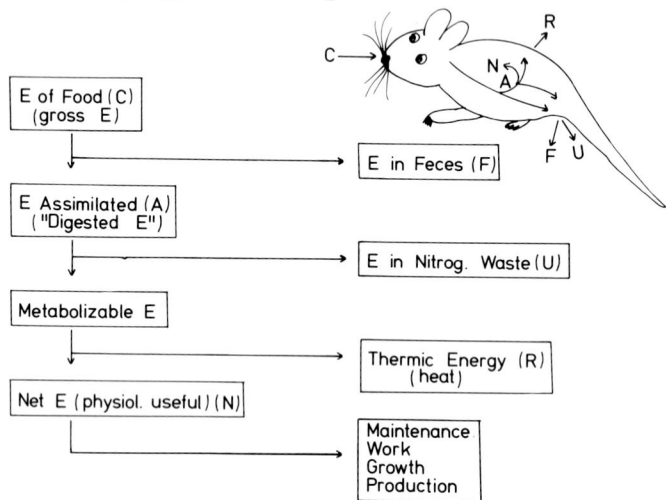

Figure 1. The fate of energy (E) in the body. Thermic energy (R) includes thermogenesis (work-induced, diet-induced, and thermoregulatory). Adapted from Brafield & Llewyllyn (1982).

The hypothesis that DR reduces free radical production by either or both of two metabolic effects (↓ rate, ↑ efficiency) is supported by increasing amounts of indirect evidence showing that long-term DR lowers tissue levels of lipid peroxides and lipofuscin (see Weindruch & Walford, 1988 [pp. 176-178]) but on nothing approaching direct proof. This hypothesis requires investigation using assays providing precise detection of the active oxygen species produced and the damage they may inflict.

The possibility that DR increases the capacity to detoxify metabolically generated active oxygen is also supported by the results of recent studies. In a collaborative study with Koizumi (Koizumi et al., 1987), we compared livers of 12 and 24 month old mice fed control (C, 95 kcal/wk) and restricted (R, 55 kcal/wk) diets since 3 weeks of age. Many enzyme activities were measured including several xenobiotic metabolizers, radical scavengers (catalase, SOD, glutathione peroxidase), and superoxide sources (xanthine oxidase, peroxisomal β-oxidation). Lipid peroxidation was also measured. The strongest dietary effect was an increased catalase activity for DR mice (group R: 42% higher at 12 months, 64% at 24 months). Lipid peroxidation was clearly lower in group R at 12 months (a 30% decrease) and somewhat lower (13%) at 24 months than in controls. These data suggest that if free radical damage is involved in aging, it may be a particular kind of damage that is in part prevented by a selective increase in catalase activity.

TABLE 2. Effects of DR and Age on Activities of Superoxide Dismutase, Catalase and on Lipid Peroxidation in Mice*

Diet	Age	SOD	Catalase	Lipid Peroxidation
C	12	43 ± 14[A]	210 ± 29[C]	27 ± 4[AB]
R	12	43 ± 8[A]	298 ± 31[B]	19 ± 2[C]
C→R	12	41 ± 3[A]	198 ± 31[C]	24 ± 3[B]
C	24	39 ± 4[A]	212 ± 15[C]	31 ± 6[A]
R	24	39 ± 7[A]	347 ± 67[A]	27 ± 5[AB]

*Ages are in months, other values are means ± SD. Abbreviations: SOD, superoxide dismutase; C, control; R, restricted; C→R, C diet until 1 week before assay when switched to R. The values for SOD (units/mg protein) and catalase (μmol/[mg protein·min]) are enzyme activities whereas those for lipid peroxidation (pmol/[mg protein·h]) give the amount of malondialdehyde formed in the TBA assay. Means in each column not sharing a common superscript letter were significantly different (P = 0.05). Adapted from Koizumi et al. (1987).

"Protein Synthesis, DNA, Chromatin, Gene Expression, DNA Repair"

The subjects subsumed under the above heading are interrelated in terms both of aging theories and experimental evidence which concerns age-related and DR-related changes. A rather large gerontologic literature has developed for each of these areas; however, there have been relatively few DR studies which have addressed these phenomena (for discussion, see: Holehan & Merry, 1986; Richardson, this volume; Weindruch & Walford, 1988). There is emerging evidence that long-term DR increases: i) rates of protein synthesis and turnover in certain tissues, ii) gene expression (but the upregulation is selective), and iii) UV-induced DNA repair. Much more intensive study of DR in relation to these areas is called for.

Neuroendocrine and Immunologic Interpretations

Neuroendocrine and immunologic interpretations carry with them the idea that in highly organized animals cellular function and survival and therefore possibly aging depend largely on the extracellular homeostatic and integrative mechanisms. This approach is different from being wholly preoccupied with events at the cellular-molecular level in individual organ systems. The possibility that DR acts via the neuroendocrine system is lucidly discussed in this volume by Meites. For a consideration of neuroendocrine and immunologic interpretation of DR's actions, see Weindruch & Walford (1988) wherein we conclude that such phenomena ".....may contribute fundamentally to the mechanism whereby DR retards aging, but the available information, while tantalizing, is insufficient and too disconnected to allow a clear formulation of possibilities."

The Basal/Induced Activity Hypothesis

Johnson et al. (1986) proposed that DR may exert its influence largely at the stage of initiation of cell replication. We have suggested that DR may act by slowing the basal turnover rate of proliferative or potentially proliferative cell populations at the same time enhancing the proliferative as well perhaps as other responses to an inductive stimulus (Walford et al., 1987). Such a condition would retard programmatic or deteriorative aging while increasing adaptability to environmental stimuli, and would accord with the observation that DR animals are both long-lived and have increased resistance to disease. Our prior studies on natural killer (NK) cells support such a concept (Weindruch et al., 1983). We found diminished NK cell activity in unstimulated DR mice, but increased activity after stimulation, compared to controls.

CONCLUSION

A major research challenge in gerontology is to elucidate the mechanisms underlying DR's many actions. Although several reasonable nonmutually exclusive possibilities exist, those approaches which emphasize processes related to energy metabolism may be particularly germane.

REFERENCES

Brafield AE, Llewyllyn MJ (1982). "Animal Energetics." Glasgow: Blackie.

Holehan AM, Merry BJ (1986). The experimental manipulation of ageing by diet. Biol Rev 61:329-368.

Johnson BC, Gajjar A, Kubo C, Good RA (1986) Calories versus protein in onset of renal disease in NZB x NZW mice. Proc Natl Acad Sci USA 83:5659-5662.

Koizumi A, Weindruch R, Walford RL (1987). Influences of dietary restriction and age on liver enzyme activities and lipid peroxidation in mice. J Nutr 117:361-367.

Masoro EJ (1988). Food restriction in rodents: An evaluation of its role in the study of aging. J Gerontol 43:B59-64.

McCarter R, Masoro EJ, Yu BP (1985). Does food restriction retard aging by reducing the metabolic rate? Am J Physiol 248:E488-490.

Walford RL, Harris SB, Weindruch R (1987). Dietary restriction and aging: Historical phases, mechanisms, current directions. J Nutr 117:1650-1654.

Weindruch R, Walford RL (1988). "The Retardation of Aging and Disease by Dietary Restriction." Springfield, IL: Charles C Thomas.

Weindruch R, Devens BH, Raff HV, Walford RL (1983). Influence of aging and diet restriction on natural killer cell activity in mice. J Immunol 130:993-996.

IMMUNE FUNCTION IN AGING RATS: EFFECTS OF GERMFREE STATUS AND CALORIC RESTRICTION

Kara W. Eberly[1] and E. Bruckner-Kardoss[2]

Saint Mary's College[1] and Lobund Laboratory[2]

Notre Dame, IN 46556

Changes in the mammalian immune system with aging have been well documented and thoroughly reviewed (Makinodan and Kay, 1980; Gilman, 1984). In healthy humans and long-lived rodent strains the decline in cell mediated immunity appears to occur earlier and more precipitously than the decline in humoral immunity (Averill and Wolf, 1985; Callard, 1978; Gillis et al., 1981). Generally numbers of T and B lymphocytes and macrophages in peripheral blood and lymphoid tissues remain constant during aging (Averill and Wolf, 1985; Gilman et al., 1981). Age-related increases and decreases of suppressor cell activity (DeKruyff et al., 1980; Ceuppens and Goodwin, 1982; Makinodan and Kay, 1980), changes in lymphocyte membrane composition and fluidity, and reduced ability to transduce stimulatory signals (Woda et al., 1979; Gilman et al., 1981; Miller, 1986; Proust et al., 1987) have been offered as explanations for the decline in cell mediated immunity. Although macrophage function has generally been found adequate in aging animals (Callard, 1978; Perkins et al., 1982; Rosenberg et al., 1983), increased prostaglandin production by macrophages (Rosenstein et al., 1980), and increased sensitivity of lymphocytes to prostaglandin inhibition (Goodwin and Messner, 1979) have also been offered as explanations.

However, it has been extremely difficult to determine if any of these processes are primary consequences of aging or secondary to other age-related alterations. Dietary restriction without malnutrition, extends lifespan and delays some age-related immune defects (Weindruch et al., 1979; Weindruch et al., 1982; Masoro, 1988). The absence

of a microbial flora, germfree status, also extends
lifespan (Gordon et al., 1966) and alters immune
parameters, such as the activation state of macrophages
(Jungi and McGregor, 1978; Mattingly et al., 1979),
development of tolerance, and response to
lipopolysaccharide (Bealmer et al., 1984; Mattingly et al.,
1979). The lifespan and pathology of the Lobund-Wistar rat
have been well characterized (see the chapters by D. Snyder
and M. Pollard, these proceedings), which makes the L-W rat
an excellent model to study the interactions of aging,
dietary restriction, and germfree status on immune
function.

Cell mediated immunity (CMI) was evaluated by the
response of splenocytes to the T-cell mitogens,
Phytohemagglutinin and Concanavalin A, and by the ability
to respond in one-way allogeneic and syngeneic mixed
lymphocyte reactions. Antibody mediated immunity (AMI) was
evaluated by the proliferative response to two
lipopolysaccharide preparations. Natural killer (NK) cell
activity was evaluated on whole spleen and nylon wool
enriched lymphocytes.

MATERIALS AND METHODS

Male Lobund-Wistar (L-W) rats were housed, fed, and
sacrificed as described by Dr. D. Snyder in this volume.
Briefly, rats were housed in flexible film isolators under
germfree (GF) or conventional (CV) conditions and fed
autoclaved Teklad L485 diet ad libitum (F) or 12 g per day
(R) beginning at 6 weeks old. This results in approximately 30% food restriction for adult animals. Rats (8)
from each group, CV-F, CV-R, GF-F, GF-R, were sacrificed at
ages 6 (young adult), 18 (middle aged), and 30 months
(old). All data were analyzed by a 3-way ANOVA from the
SAS statistical package, and statistical significance is
$p<.01$ unless otherwise noted.

Orbital blood for peripheral white cell and differential counts was obtained under halothane anesthesia.
Spleen cells for functional assays were aseptically dis-
aggregated between frosted microscope slides into Hanks
Balanced Salt Solution, washed twice, counted and diluted
into RPMI 1640 medium supplemented with 5% fetal calf
serum, antibiotics, glutamine, pyruvate, MEM nonessential

amino acids, and 2-mercaptoethanol (Mishell and Shiigi, 1980). Cell viability was consistently 90% or better by trypan blue exclusion.

For mitogen assays spleen cells (2×10^5/ well) were incubated with mitogen in 200 mcl supplemented RPMI 1640 at 37C in 5% CO_2 in microtiter plates. Tritiated thymidine (1 mcCi/well) was added the last 16 hours of a 3 day assay; a Mini-Mash cell harvester was used; and activity was counted in a liquid scintillation counter. Results were expressed as counts per minute (cpm) with mitogen minus cpm without mitogen. Phytohemagglutinin-P (Difco) was used at 23 mcg/ml and Concanavalin A (Pharmacia) at 1.3 mcg/ml. The two lipopolysaccharide preparations, STM from Salmonella typhimurium (Ribi) and LPS from Salmonella typhosa (Difco) were used at 25 mcg/ml, which was optimal for LPS and suboptimal for STM.

Stimulators for the mixed lymphocyte reactions were spleen cells from 6 month LW rats (SMLR) or 3 month Lobund Sprague-Dawley rats (MLR) treated with mitomycin C and diluted to 2×10^5 cells/well. The assay was conducted as the mitogen assay except the label was added the last 16 hours of a 5 day assay (Mishell and Shiigi, 1980).

NK activity was assayed in a 4-hour ^{51}Cr release assay against YAC-1 mouse lymphoma cells in ratio of 100:1 for whole and nylon wool enriched spleen cells (Tazume and Pollard, 1985). Daily variation in the assay was corrected by including 3 month Sprague-Dawley females in each assay and determining a correction factor which brought the percent lysis of that animal to the level of the mean of all SD females. That correction factor was then used for all data in that experiment (Rusthoven J, 1985).

RESULTS

Younger and fullfed rats had slightly higher peripheral white cell counts, but the differences were not significant. The percentage of lymphocytes decreased and neutrophils increased with age in all groups except GF-R. Floral status was insignificant, and diet was only important at 18 months when R rats had more lymphocytes than F. The numbers of lymphocytes per mm^3 peripheral blood declined significantly between 6 and 18 months; both

F (p<.01) and CV (p<.05) rats had higher counts. Neutrophil counts increased significantly only after 18 months and neither floral status nor diet was significant. Although the percentage composition of peripheral blood in 18 month CV-R rats was more like 6 months CV rats, the actual lymphocyte count was similar to 30 month CV-F.

Peripheral white cell counts were higher in F than R rats, but floral status was not significant. Only CV-F rats showed a significant (p<.05) decline with age. However, the percentage of lymphocytes did decline significantly with age in all groups except the GF-R rats, and the percentage of neutrophils rose in a compensatory fashion. Floral status did not affect the peripheral blood neutrophil and lymphocyte composition, but restriction slowed the change in CV rats.

Thymus size declined significantly between 6 and 18 months, F animals had larger thymuses than R, but floral status was not significant. When thymus size was expressed as percent body weight, only the age-related changes were significant. Spleen weight was independent of floral status, but F rats had heavier spleens than R. Spleen size increased significantly with age only among R rats. When spleen size was expressed as percent body weight, R rats had higher values than F.

Splenic natural killer NK activity increased between 6 and 18 months of age; GF rats had higher activities than CV; and F was higher than R (Figure 1). Nylon wool enriched NK activity also increased between 6 and 18 months; but difference due to floral status disappeared; and the difference between F and R was not as pronounced (p<.05). When the data were examined without the correction for day to day variation, the age difference remained, but differences due to floral and dietary status were not significant.

Optimal mitogen concentrations did not differ with age, diet, or floral status. Proliferative responses to the T-cell mitogens, PHA and Con A, declined significantly between 6 and 18 months, and responses were higher in CV than GF rats (p<.05). Dietary restriction resulted in significantly higher responses to PHA but not to Con A (Figure 2). However, among CV rats, the response to both mitogens was significantly higher in R than F rats at 18 months. The 18 month response was not significantly different from 30 month in F rats. Therefore, R did not prevent the age-related decline in T-mitogen response among CV rats, but it significantly delayed the decline.

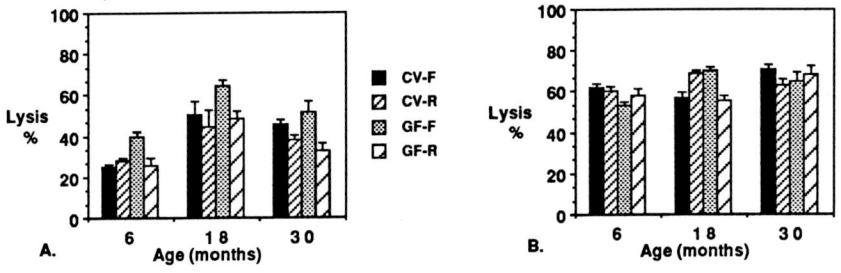

Figure 1. Natural killer activity of splenocytes from L-W rats. The effector to target ratio was 100:1. Bars represent the mean, and error bars are the standard error of the mean. A. Whole spleen, B. Nylon wool enriched spleen cells.

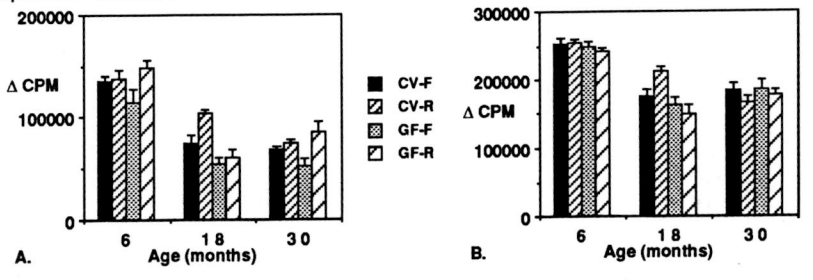

Figure 2. Proliferative response of spleen cells from L-W rats to the T-cell mitogens Phytohemagglutinin and Concanavalin A. Bars represent the mean, and error bars are the standard error of the mean. A. PHA, B. Con A.

GF rats responded more strongly to both lipopolysaccharide preparations, Difco LPS and Ribi STM; and there was a significant age-related decline in both responses. F rats responded more strongly to LPS, but not to STM (Figure 3). There was also an age-related decline in the response to pokeweed mitogen; CV responses were higher than GF, but diet was not significant.

Both the (SMLR) and the allogeneic MLR declined significantly with age. The SMLR was very low in all cases, and neither floral nor dietary status were significant. Restricted rats gave a higher response in the

MLR, and floral status was not significant (Figure 4).

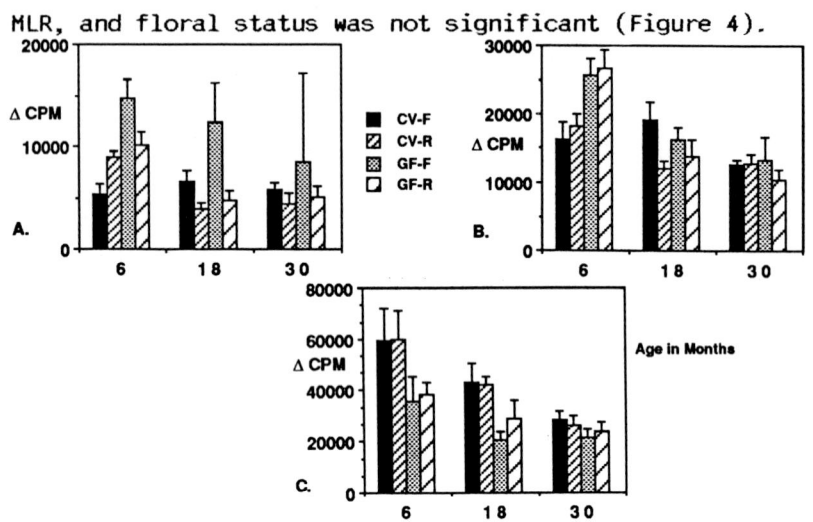

Figure 3. Proliferative response of spleen cells from L-W rats to two lipopolysaccharide preparations and pokeweed mitogen. A. Difco LPS, B. Ribi STM, C. PWM

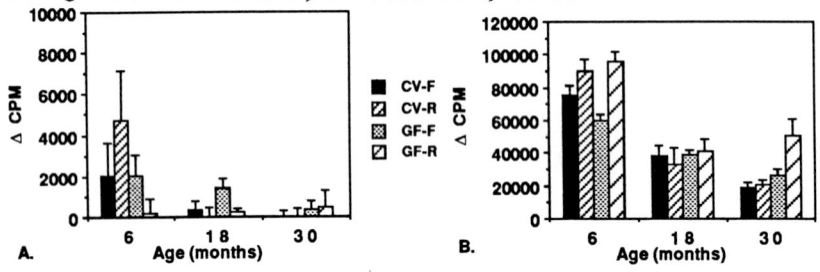

Figure 4. Proliferative response of spleen cells from L-W rats to A. syngeneic spleen cells (SMLR) and B. allogeneic spleen cells (MLR).

DISCUSSION

This work examined the effects of mild food restriction, instituted gradually at 6 weeks of age, and the absence of a normal microflora, on several immune parameters in young adult (6 month), middle aged (18 month), and old (30 month) LW rats. Preliminary work established that the immune parameters, although changing between 3 and 5 months, were stable from 5 to 8 months of age, so 6 months was chosen for the young adult sample.

Early mortality in CV-F LW rats began shortly after 18 months, but very little pathology was found at that age. Thirty months is close to the median survival for this strain. Parameters for evaluation were chosen based on the following criteria: 1) reported to vary with age, 2) represented cell-mediated, antibody-mediated, or non-specific responses, and 3) were entirely noninvasive so that other aspects of the rat's physiology would not be altered.

Peripheral blood lymphocytes (PBL) have been reported to decrease or remain the same with age (Kishimoto et al., 1978; Kay, 1979). Lymphocytes declined while neutrophils increased with age in L-W rats, and immune stimulation by a conventional microflora and ad libitum feeding were both associated with higher lymphocyte counts. Surprisingly, floral status had no effect on neutrophil counts, perhaps because the LW rats were GF-derived and maintained under very clean conditions. Young CV rats had larger thymuses than GF, which confirmed previous reports (Bealmer et al., 1984). When differences in total body weight were taken into account, only age significantly affected thymus size (young>old); and only diet affected spleen size (R>F).

Natural killer activity can be considered a nonspecific defense against virus infection and cancer (Barlozzari et al., 1983), particularly hematopoietic cancers; and rat splenic NK activity has been reported to decline with age (Bash and Vogel, 1984) Although L-W rats did not have a high incidence of lymphoid cancers, a considerable amount of malignant and premalignant pathology developed with age. In the L-W rat splenic NK activity increased between 6 and 18 months. GF and F status both enhanced NK activity in whole spleen preparations, but nylon wool enrichment eliminated the floral difference and decreased the dietary effect. This supports the report of Weindruch et al (1983) on the depressing effect of R on unstimulated NK activity in mice. Using nylon wool enriched spleen cells, Tazume and Pollard (1985) reported that GF Sprague-Dawley rats had higher NK activity than CV, but in the L-W rat nylon wool enrichment eliminated the floral effect. Perhaps the CV spleen contains more B-cells which are eliminated by nylon wool thus causing more enrichment of CV NK activity. These data do not support a role for deficient NK activity in the development of neoplastic disease in the LW rat.

CMI has been reported to decline earlier and more precipitously than AMI (Kay, 1979; Makinodan and Kay, 1980; Gilman, 1984; Gillis et al., 1981), and dietary restriction slowed the age-related CMI decline (Weindruch et al.,1979; Weindruch et al., 1982). The L-W rat developed T cell defects with aging which to some extent were overcome by dietary restriction. Optimal mitogen concentrations did not differ with age, diet, or floral status. T-cell mitogen responses declined most rapidly between 6 and 18 months, and the PHA decline was more pronounced than Con A. Dietary restriction caused the greatest enhancement on PHA responses of middle aged CV rats, although it also enhanced the Con A response at that age. Therefore, R slowed the age-related decline to both T-cell mitogens. CV rats gave moderately higher responses to both T-cell mitogens than GF.

The syngeneic and allogeneic mixed lymphocyte reactions are both responses to self and foreign Class II antigens on non-T cells (Salomon et al., 1983). Young stimulator cells were used because old macrophages are not as effective as stimulators (Callard, 1978). In the SMLR, T-cells secrete IL3, but not IL2 nor IFN. The IL3 in turn stimulates the proliferation of early precursors in lymphocyte or hematopoietic cell differentiation. In the MLR, IL2, IL3 and IFN are produced, and there is proliferation of the same cells as in the SMLR, as well as mature T cells and NK cells which respond to IL2 and IFN. Cytotoxic T cells are also generated (Suzuki et al., 1986). The L-W SMLR was very low throughout, declined with age, and was unaffected by floral or dietary status. The MLR also declined with age, and R rats generally had higher responses than F. However, in this case the difference was more dramatic at the younger rather than older ages. Floral status had no effect.

Responses to pokeweed mitogen (PWM) are believed to be chiefly due to B-cells that respond to T-dependent antigens, but the response is also modulated by T-helper and T-suppressor cells (Ceuppens and Goodwin, 1982; Kay, 1979). There was an age-related decline in the L-W rat response to PWM; diet was irrelevant; and CV rats responded more actively than GF. This probably reflects alterations in spleen B- and T-cell populations in response to antigenic stimulation. Although there are dead microbes in the diet, the response of a GF animal is much more likely to be a true primary response; this is supported by the much lower serum immunoglobulin levels in GF rats (Bealmer et al., 1984).

Aging immune responses to lipopolysaccharide (LPS) have been reported to increase, decrease, or remain the same (Makinodan and Kay, 1980), while GF animals gave an enhanced response (McGhee et al., 1980). Although rats do not make a very dramatic proliferative response to LPS, the absence of LPS in the environment results in considerable differences in splenic and peritoneal macrophages. CV rat macrophages produce considerable prostaglandin E2, and are therefore quite immunosuppressive in several assays (Rosenstein and Strauser, 1980; Rosenberg et al., 1983). Germfree macrophages are at a lower activation state (Jungi et al., 1978), produce less prostaglandin E2, and are much less suppressive (Mattingly et al., 1979). Aged CV rat macrophages appear to have an increased activation state which is associated with less efficient presentation of antigen in some situations, and therefore a less efficient immune response (Lovik et al., 1985; Orme, 1987). Therefore, the GF L-W rat allows the examination of aging in the absence of LPS generated macrophage suppression.

Aging resulted in decreased proliferative responses to both Difco LPS and STM. F rats responded more strongly to Difco LPS but not STM, which indicates that the nature of the LPS preparation is important. GF status enhanced responses to both preparations. The age-related decline in both GF and CV rat responses supports a B-cell defect with aging, while the greater GF response supports the role of suppression in the LPS response in CV animals.

In conclusion, the L-W rat appeared to develop T cell defects with aging which to some extent were overcome by dietary restriction. However, germfree status had either no effect (SMLR or MLR), or led to a reduced response (PHA, Con A). We also presented evidence for B cell defects in the responses to PWM and two lipopolysaccharide preparations. However, dietary restriction had no effect on PWM response and decreased the response to one of the LPS preparations. GF status led to a decreased response to PWM, probably due to lack of previous antigenic stimulation. However, GF status enhanced the response to LPS, probably because of a lack of macrophage suppressors (Mattingly et al., 1979). Finally, there was no age-related decline in NK activity beyond 6 months, and neither floral nor dietary status had dramatic effects. Therefore, it is unlikely that changes in NK activity play a role in the development of malignancy in the LW rat.

REFERENCES

Averill L, Wolf N (1985). The decline in murine splenic PHA and LPS responsiveness with age is primarily due to an intrinsic mechanism. J Immunol 134:3859-3863.

Barlozzari T, Reynolds CW, Herberman RB (1983). In vivo role of natural killer cells. Involvement of LGL in the clearance of tumor cells in anti-asialo GM1 treated rats. J Immunol 131:1024-1027.

Bash JA, Vogel D (1984). Cellular immunosenescence in F344 rats: Decreased natural killer cell (NK) activity involves changes in regulatory interactions between NK cells, interferon, prostaglandins, and macrophages. Mech Ageing Dev 24:49-65.

Bealmer PM, Holtermann OA, Mirand EA (1984). Influence of the microflora on the immune response. In Coates ME, Gustafson BE (eds): "The Germfree Animal in Biomedical Research," London: Laboratory Animals Ltd, pp 335-384.

Callard R (1978). Immune function in aged mice III. Role of macrophages and effect of 2-mercaptoethanol in response of spleen cells from old mice to phytohaemagglutinin, lipopolysaccharide and allogeneic cells. Euro J Immunol 18:697-705.

Ceuppens JL, Goodwin JS 1982. Regulation of immunoglobulin production in pokeweed mitogen-stimulated cultures of lymphocytes from young and old adults. J Immunol 128:2429-2434.

DeKruyff RH, Kim YT, Siskind GW, Weksler ME (1980). Age-related changes in the in vitro immune response: Increased suppressor activity in immature and aged mice. J Immunol 125:142-147.

Gillis S, Kozak R, Durante M, and Weksler ME (1981). Immunological studies of aging: decreased production of and response to T cell growth factor by lymphocytes from aged humans. J Clin Invest 67:937-942.

Gilman SC, Woda BA, Feldman JD (1981). T lymphocytes of young and aged rats. I. Distribution, density, and capping of T antigens. J Immunol 127:149-153.

Gilman S (1984). Lymphokines in immunological aging. Lymphokine research 3:119-123.

Goodwin JS, Messner RP (1979). Sensitivity of lymphocytes to prostaglandin E2 increases in subjects over age 70. J Clin Invest 64:434-439.
Gordon HA, Bruckner-Kardoss E, Wostmann BS (1966). Aging in germfree-mice: life tables and lesions observed at natural death. J Gerontol 121:380-387.
Jungi TW McGregor DD (1978). Impaired chemotactic responsiveness of macrophages from gnotobiotic rats. Infect Immun 19:553-561.
Kay MMB (1979). An overview of immune aging. Mech Ageing Dev 9:39-59.
Kishimoto S, Tomino S, Inamata K, Kotegawa S, Saito T, Kuroki M, Mitsuya H, Hisamitsu S (1978). Age-related changes in the subsets and functions of human T lymphocytes. J Immunol 121:1773-1775.
Makinodan T, Kay M (1980). Age influence on the immune system. Adv Immunol 29:287-330.
Masoro EJ (1988). Minireview: Food Restriction in rodents: An evaluation of its role in the study of aging. J. Gerontol 43:B59-64.
Mattingly JA., Eardley DD, Kemp JD, Gershon RK (1979). Induction of suppressor cells in rat spleen: Influence of microbial stimulation. J Immunol 122:787-790.
McGhee JR, Kiyono H, Michalek SM, Babb JL, Rosenstreich DL, Mergenhagen SE (1980). Lipopolysaccharide (LPS) regulation of the immune response. T lymphocytes from normal mice suppress mitogenic and immunogenic responses to LPS. J Immunol 124:1603-1611.
Miller RA (1986). Immunodeficiency of aging: restorative effects of phorbol ester combined with calcium ionophore. J Immunol 137:805-808.
Mishell BB and Shiigi SM (1980). "Selected Methods in Cellular Immunology". San Francisco: W H Freeman, pp 30-37 and 162-164.
Orme IM (1987). Aging and immunity to tuberculosis: increased susceptibility of old mice reflects a decreased capacity to generate mediator T lymphocytes. J Immunol 138:4414-4418.
Perkins EH, Massuci JM, Gloven PL (1982). Antigen presentation by peritoneal macrophages from young adult and old mice. Cell Immunol 70:1-10
Proust JJ, Filburn CR, Harrison SA, Buchholz MA, Nordin AA (1987). Age-related defect in signal transduction during lectin activation of murine T lymphocytes. J Immunol 139:1472-1478.
Rosenberg JS, Gilman SC, Feldman JD (1983). Effects of

aging on cell cooperation and lymphocyte responsiveness to cytokines. J Immunol 130:1754-1758.

Rosenstein MM Strauser HR (1980). Macrophage-induced T cell mitogen suppression with age. J Reticuloendothel Soc 27:159-166.

Rusthoven J (1985). Clinical significance of natural killer cell cytotoxicity: Need for proper data analysis in the design of clinical studies. J Cancer Clin Oncol 21:1287-1293.

Salomon DR, Cohen DJ, Carpenter CB, Milford EL (1983). Regulation of the immune response to alloantigens: suppressor and helper T cells generated in the primary MLR of the rat. J Immunol 131:1065-1072.

Tazume S, Wade A, Pollard M (1985). Natural killer activity in germfree and conventional rats. Prog Clin Biol Res 181:343-345.

Weindruch, R, SRS Gottesman, and RL Walford. 1982. Modulation of age-related immune decline in mice dietarily restricted from or after midadulthood PNAS 79:898-902

Weindruch RH, Kristie JA, Cheney KE, Walford RL (1979). Influence of controlled dietary restriction on immunologic function and aging. Fed Proc 38:2007-2016.

Weindruch RH, Devens BH, Raff HV, Walford RL (1983). Influence of aging and diet restriction on natural killer cell activity in mice. J Immunol 130:993-996.

Woda BA. Yguerabide J, Feldman JD (1979). Mobility and density of AgB, "Ia" and Fc receptors on the surface of lymphocytes from young and old rats. J Immunol 123:2161-2167.

EFFECT OF DIETARY RESTRICTION AND AGING ON LYMPHOCYTE SUBSETS IN GERMFREE AND CONVENTIONAL LOBUND-WISTAR RATS

Yoon Berm Kim and Alice Gilman-Sachs

Department of Microbiology and Immunology, University of Health Sciences/The Chicago Medical School, 3333 Green Bay Road, North Chicago, Illinois 60064

INTRODUCTION

Aging in normal individuals is accompanied by a loss in the ability to respond to environmental stress and alterations in the immune system (Makinodan and Kay, 1980). This process is accompanied primarily by a decline in T cell function, perhaps due to the thymic atrophy that begins with the onset of sexual maturity. Thus, in vivo immune responses such as delayed hypersensitivity to primary antigens decline in the elderly and the in vitro proliferative capacity of T cells from humans and rodents in response to PHA, concanavalin A and allogeneic target cells decreases with age (Siskind, 1987). In contrast the total number of B cells does not change appreciably with age under most conditions although there may be a small but statistically significant change in the concentration of serum immunoglobulins with age. Also, the immune response to environmental antigen tends to decrease and a rise in the production of auto-antibody is observed with age. Several studies have shown that dietary restriction extends the life span of rats and mice and alters the rate of age-related changes in immunological parameters (Weindruch et al., 1982). Dietary manipulations produce striking influences on immunologic function, disease manifestations, and longevity in autoimmune-prone mice (Fernandes et al., 1976; Kubo et al., 1984a,b). In (NZB/NZW)F1 mice decreased calorie intake prolonged life and greatly decreased production of anti-DNA antibodies and increased IL-2 production (Kubo et al., 1984 a). In MRL/Mp-1pr mice restriction of food intake inhibited development of lymphoproliferative disease and greatly decreased numbers of cells in thymus, lymph nodes and spleen (Kubo et al., 1984b). One way to evaluate the immune system is to determine the number of lymphocytes in peripheral blood and the subset composition. A comparison of these immune parameters in young and old individuals can give an indication of the effect of the aging process and diet on these parameters.

The Lobund-Wistar Rat (L-W) represents a unique opportunity to study the effect of diet and environmental antigens on the development of the immune response. The life spans of these rats can be extended experimentally by two interventions known to increase life expectancy in rats: life in a germfree environment and restricted dietary intake (Pollard and Wostmann, 1985; Snyder and Wostmann, 1987). In L-W rats older than 30 months of age, there is the consistent increase of tumors. This finding may represent a defect in the immune system or may represent other manifestations due to aging. Therefore, the effect of dietary restriction and environmental antigens on the immune system was investigated by determining the expression of lymphocyte subsets in peripheral blood of four groups of L-W rats over a period of 30 months (6 months, 18 months, and 30 months sampling periods).

MATERIALS AND METHODS

Heparinized blood was obtained by cardiac puncture from each of 4 groups of rats raised under different environmental conditions at intervals of 6 months, 18 months and 30 months of age (see chapter by Snyder and Wostmann in these proceedings). Fifty to 100 ul of whole blood containing approximately 2×10^6 lymphocytes were reacted with a monoclonal antibody to a rat lymphocyte subset marker and analyzed by flow cytometry to determine the percentage of peripheral blood lymphocytes expressing these markers. White cell counts were determined in a Coulter ZBI Counter; differential blood cell counts were determined after staining with Fisher Leukostain. The number of lymphocytes per ml were determined by multiplying the white blood cell count by the percentage of lymphocytes.

Monoclonal Antibodies. The monoclonal antibodies used in these studies were W3/13 (pan T cells and NK cells), OX19 (peripheral T cells), W3/25 (T-helper cells), Ox8 (T-suppressor/cytotoxic cells and natural killer cells), OX4 and OX6 (monomorphic determinants of Ia molecules) (Mason et al., 1983) (Serotec/Bioproducts for Science). Peripheral blood containing at least 5000 lymphocytes was reacted with a volume of ascites fluid or tissue culture supernatent fluid determined to be in excess, incubated for 30 min on ice at 4° C, and washed three times at 1200 rpm for 5 min. Fluorescein-conjugated F(ab)'2 goat anti-mouse Ig with no cross reaction to rat Ig (Cappel Labs) was then added (1:10 dilution, 25 ul). Each tube was incubated for 30 min at 4° C and washed three times. Red blood cells were then lyzed with Coulter Immunolyze (Coulter Immunology, Hialeah, FL), (0.5 ml of 1:25 dilution) and fixed with Coulter Fixative (0.125 ml/tube) (Thompson et al., 1986). After washing twice more, the stained cells were analyzed by flow cytometry.

For two color fluorescence, blood was reacted first as described above with monoclonal antibody, incubated with 25 ul of a 1:5 dilution of mouse serum, washed, and then reacted with biotinylated monoclonal antibody previously isolated by ammonium sulfate precipitation (40%). The antibody was then biotinylated. The biotinylated monoclonal antibody was detected with avidin-phycoerthyrin (Fisher). All reagents were determined to be antibody excess in preliminary experiments. Controls consisted of mouse Ig and the same indirect reagents.

Flow Cytometric Analysis. Fluorescent lymphocytes were analyzed on a Coulter EPICS 751 Flow Cytometer using excitation with the 488 nm lines of an argon laser. Lymphocytes were gated based on 90° light scatter and forward angle light scatter. This population was then analyzed for the percentage of lymphocytes positive for FITC (green fluorescence) by collecting the fluorescence at 530 nm. For two color fluorescence, green fluorescence and red fluorescence (collected at 570 nm) were determined. Electronic subtraction of green into red fluorescence was performed when fluorescein and phycoerythrin were used in combinations. Data analysis was performed by using the MDADS computer programs. Background subtraction was performed for each specimen.

RESULTS

Peripheral blood was analyzed by flow cytometry to determine the percentages of lymphocyte subsets in four groups of rats raised under different environmental conditions over a period of 30 months. Samples were obtained at 6, 18 and 30 months, stained with monoclonal antibodies to lymphocyte subsets, and analyzed by flow cytometry to determine what significant differences could be found between the percentages of lymphocyte subsets in the four groups at the 3 different ages. The four groups of rats were raised under: 1) conventional conditions and allowed to feed freely (CV-F); 2) conventional conditions and placed on a restricted diet (CV-R); 3) germfree conditions and allowed to feed freely (GF-F); and 4) germfree conditions and placed on a restricted diet (GF-R) and were maintained at the Lobund Laboratory, University of Notre Dame.

T lymphocytes were identified with monoclonal antibody to OX19, a pan-T cell marker. $OX19^+$ T lymphocytes were found at a higher percentage for all groups at 6 months than at 30 months (Fig. 1). Thus, for the CV-F group at 6 months, the average percentage was 49.2% which declined to 35.2% at 30 months; for the CV-R group, the average percentage at 6 months was 60.4% which declined to 38.7% at 30 months; for the GF-R group, the average percentage at 6 months was 58.3% which declined to 35.2% at 30 months; and for the GF-R group the average percentage was 59.9% which declined to 25.8% at 30 months. There was no statistical difference between the percentage of OX19 bearing cells in all groups at 6 months, 18 months, or 30 months; there was no difference between the GF and CV groups. Thus, there appeared to be a decline in the expression of the OX19 marker on T lymphocytes in 30 month old rats in contrast to the expression of W3/13 marker on T lymphocytes.

Natural killer cells were identified by two color fluorescence using a combination of anti-OX8 (FITC) and anti-OX19 (PE). It has previously been shown that NK cells are $OX8^+/OX19^-$ lymphocytes (Woda et al., 1986). With regard to age, the percentage of NK cells was less at 6 months than at 30 months for three of the four groups (Fig. 2). The only exception was the GF-R group where the percentage at 6 months was similar to the percentage at 30 months with regard to diet; the restricted diet groups have a higher percentage at 30 months of age than that at 6 months compared to the full fed group. Thus, there appeared to be an effect of diet on expression NK cells.

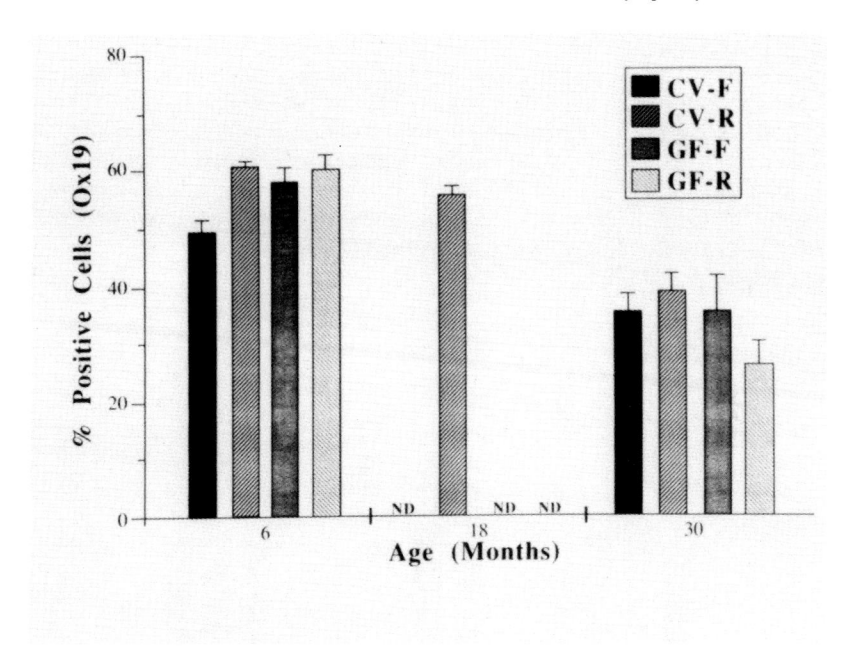

Fig. 1 Total T lymphocytes in 4 groups of rats

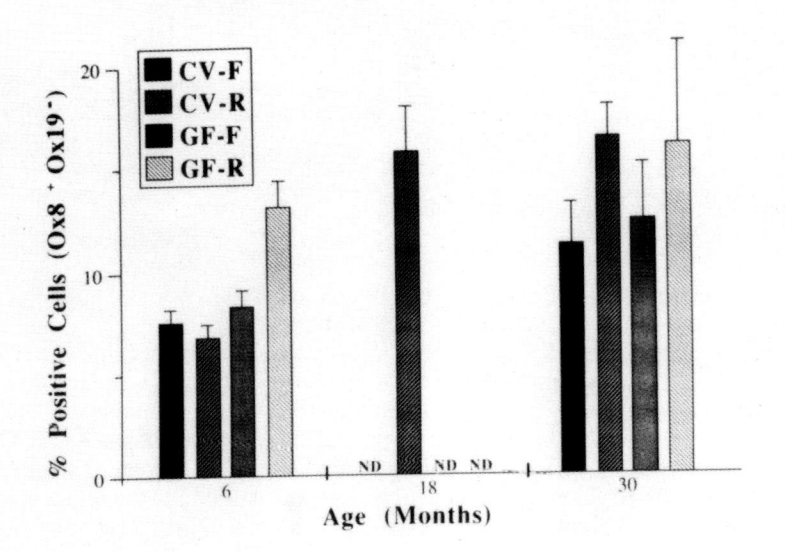

Fig. 2 NK cells in 4 groups of rats

In the rat, two monoclonal antibodies identify non-overlapping T cell populations responsible for either helper or suppressor activity. These two monoclonal antibodies, W3/25, specific for T-helper lymphocytes and OX8, specific for T-suppressor/cytotoxic lymphocytes (Ts/c) and NK cells, were used to analyze the T-cell subpopulations. For all four groups of rats at the three different ages, there were no significant differences in the expression of W3/25 (T-helper cell marker) (Fig. 3).

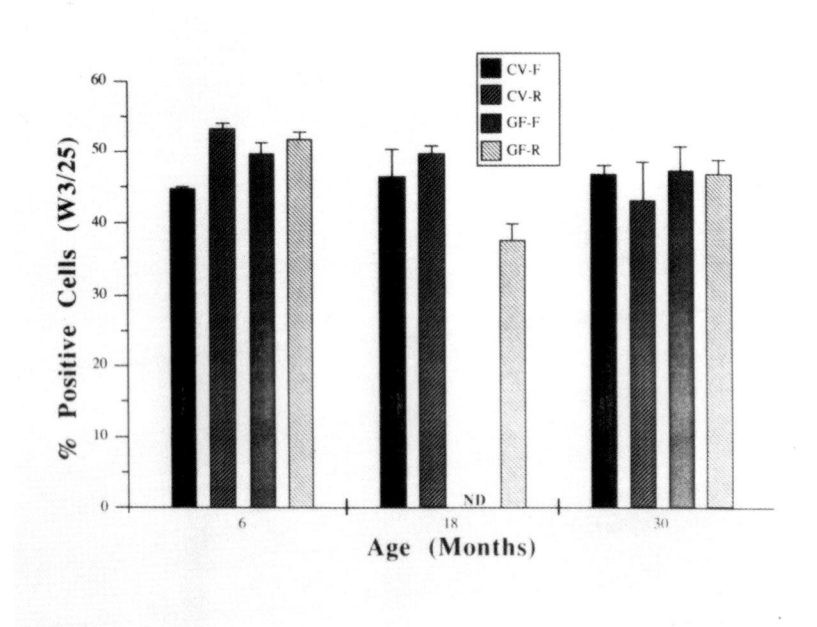

Fig. 3 T helper cells in 4 groups of rats

In addition to reacting with Ts/C lymphocytes, OX8 reacts also with natural killer cells. Therefore an accurate estimate of Ts/c lymphocytes would be the percentage of $OX8^+ - OX8^+/OX19^-$ (NK) lymphocytes (Fig. 4). Based on this analysis, there were more Ts/c lymhocytes in germfree than conventional groups and more Ts/c lymphocytes in fullfed groups than restricted groups. Also, there was an age effect; there was more Ts/c at 6 months in all groups than 30 months.

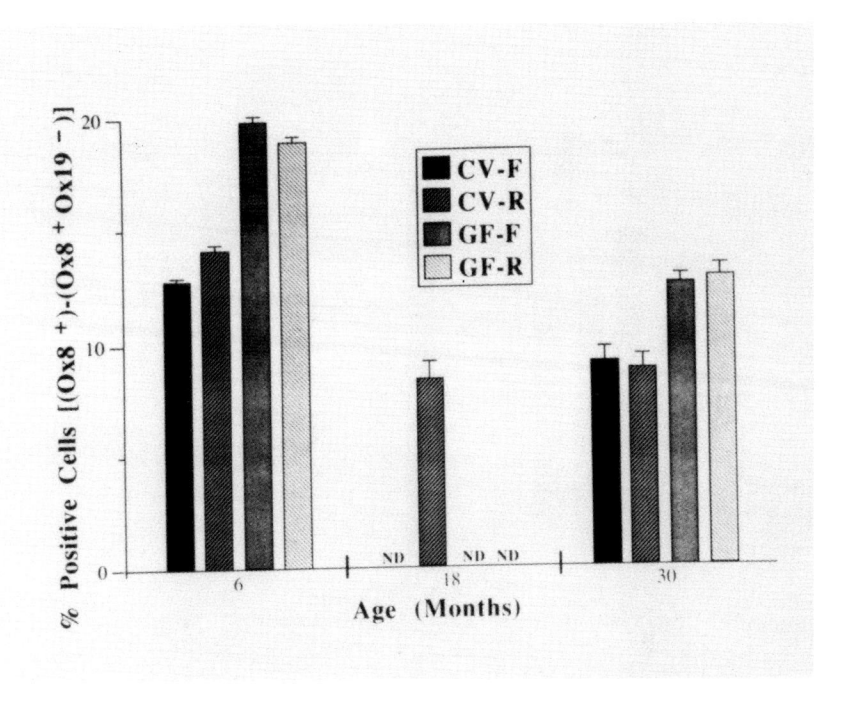

Fig. 4 T cytotoxic/suppressor cells in 4 groups of rats

Monoclonal antibodies to monomorphic determinants on Ia molecules (OX6 and OX4) were used to identify B cells. Monocytes that do react with these monoclonal antibodies were not included in the lymphocyte map. Similar results were found for each monoclonal antibody. There were more B lymphocytes in the conventional rats than germfree rats and more B lymphocytes in the fullfed rats than the restricted rats (Fig. 5). This finding would be consistent with the increased exposure to antigenic stimulation by environmental antigens in the conventional and full fed groups of rats. Thus at 6 months, the most B cells (OX6$^+$) were observed in the CV-F (37.2%) and the least in the GF-R (23.0%). Also, the percentage of B cells (OX6$^+$) in the CV-F group (37.2%) was greater than in the CV-R group (30.3%) and the percentage of B cells in the GF-F group (30.7%) was greater than in GF-R group (23.0%). Similar results were found in blood obtained at 30 months (i.e. CV-F > CV-R > GF-F > GF-R).

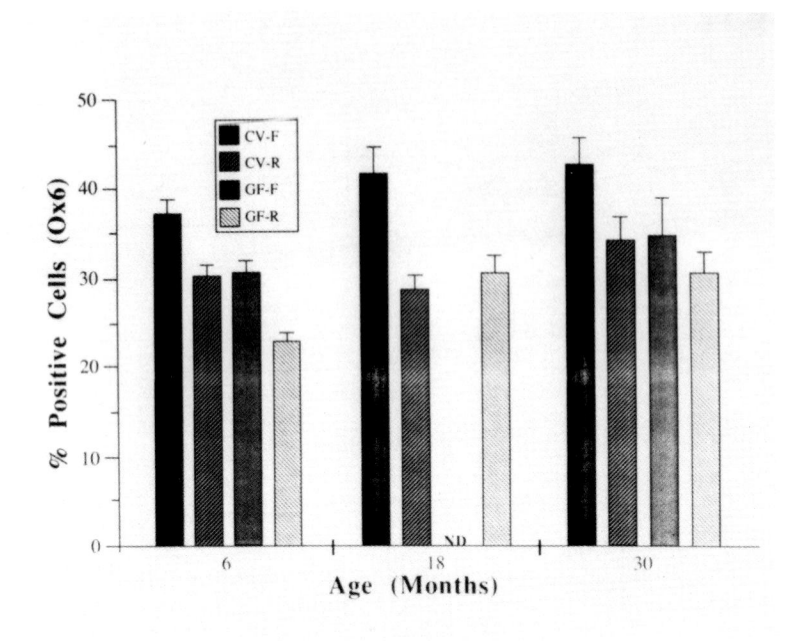

Fig. 5 B lymphocytes in 4 groups of rats

Within each group no difference in the expression of OX6 or OX4 was observed with age with the exception of the GF-R group. In this group, there was an increase in the percentage of B (OX6$^+$) cells at 18 and 30 months compared to the percentage at 6 months.

DISCUSSION

An analysis of lymphocyte subsets in 4 groups of rats raised under different dietary and environmental conditions indicated the following differences. First, the percentages of T lymphocytes as identified by OX19 was higher at 6 months than at 30 months in all groups of rats. Second, the percentage of B lymphocytes (OX6) was higher in conventionally raised rats than germfree rats and higher in fullfed rats than the group on a restricted diet. Third, natural killer cells (OX8$^+$ OX19$^-$) increased with age and were significantly higher in germfree restricted groups than other groups at 6 months of age. Fourth, there were no significant differences in T-helper cells in all groups at all ages. Fifth, T cytotoxic/suppressor cells [OX8$^+$ minus OX8$^+$ OX19$^-$ (NK cells)] were higher in germfree than in conventional animals and higher at 6 months than at 30 months.

It has been shown that with increasing age a variety of changes occurs in the immune response in many species including man. These changes tend to affect T cell dependent functions more than T-cell-independent functions (Makinodian and Kay, 1980). Several studies in humans using monoclonal antibodies to peripheral T lymphocytes (T3) have shown an age-related decline in T3-cell populations in older persons (Schwab et al., 1983). There is also a decrease in the percentage of thymic lymphocytes that express the sheep erythrocyte receptor, a marker on mature thymocytes, reflecting the decreased capacity of the thymus gland to modulate the differentiation of immature lymphocytes.

Gilman et al (1981) studied the distribution, density and capping of T cell antigens W3/13, W3/25, and Thy-1 in the rat and found that the total numbers of W3/13, W3/25 and Thy-1 positive cells were reduced in spleens and lymph nodes of aged rats. They also showed that the density of these antigens and their rate of capping was decreased in "old" cells suggesting that the membrane composition and cytoskeleton were altered in aged rats.

In addition to the change in the proportion of lymphocyte populations with age, there is also a rise in the concentration of IgA and IgG in human serum and a decrease in IgM with age (Siskind, 1987). Autoantibodies to nucleic acids, smooth muscle, mitochondria, lymphocytes, gastric parietal cells, immunoglobulins and thyroglobulin have all been found increased in older persons. Similarly, activation of B lymphocytes with a polyclonal B cell activator showed there was more autoantibody production in "older" lymphocytes than "young" lymphocytes. In addition, older animals have been shown to produce excessive auto-anti-idiotype antibody during the immune response.

ACKNOWLEDGMENTS

We wish to thank Drs. Morris Pollard, David Synder and Bernard Wostmann for their generous offer of invaluable blood samples of aging Lobund-Wistar rats and an invitation to participate in the collaborative research project.

REFERENCES

Fernandes G, Yunis EJ, Good RA (1976). Influence of protein restriction on immune functions in NZB mice. J. Immunol 116:782-790.

Gilman SC, Woda BA, Feldman J (1981). T lymphocytes of Young and Aged Rats. 1. Distribution, density and capping of T antigens. J. Immunol 127:149-153.

Kubo C, Johnson BC, Day NK, Good RA (1984a) Calorie source, calorie restriction, immunity and aging of (NZB/NZ10)F1 mice. J. Nutr 114:1884-1899.

Kubo C, Day NK, Good RA (1984b). Influence of early or late dietary restriction on life span and immunological parameters in MRL/Mp-lpr/lpr mice. Proc Natl Acad Sci 81:5831-5835.

Makinodan T, Kay MB (1980). Age influence on the immune system. Adv Immunol 29:287-330.

Mason DW, Arthur RP, Dallman MJ, Green JR, Spickett GP, Thomas ML (1983). Functions of rat T-lymphocyte subsets isolated by means of monoclonal antibodies. Immunol Rev 74:1-82.

Pollard M, Wostmann BS (1985). Aging in germfree rats: the relationship to the environment, diseases of endogenous origin and to dietary modification. In Archibald J, Ditchfield J, Rowsell HC (eds): "Contribution of Laboratory Animal Science to Welfare of Man and Animals" New York, Gustav Fischer Verlag, pp 181-186.

Schwab R, Starano-Coico L, Weksler ME (1983). Immunological studies of aging. 9. quantitative differences in T lymphocyte subsets in young and old individuals. Diagn Immunol 1:195-198.

Siskind GW (1987). Aging and the immune system. Aging 31:235-242.

Snyder DL, Wostmann BS (1987). Growth rate of male germfree Wistar Rats fed ad libitum or restricted natural ingredient diet. Lab Animal Sci 37:320-325.

Thompson SC, Bowen KM, Burton RC (1986). Sequential monitoring of peripheral blood lymphocyte subsets in rats. Cytometry 7:184-193.

Weindrich R, Gottesman S, Walford R (1982). Modification of age-related immune decline in mice dietarily restricted from or after midadulthood. Proc Natl Acad Science USA 79:898-902.

Woda BA, Like AA, Padden C, McFadden ML (1986). Deficiency of T-cytotoxic-suppressor cells in the BB/W rat. J. Immunol 136:856-859.

C-REACTIVE PROTEIN IN AGING LOBUND WISTAR RATS

Joan N. Siegel and Henry Gewurz

Department of Immunology/Microbiology
Rush-Presbyterian-St. Luke's Medical Center
Chicago, Illinois 60612

C-reactive protein (CRP) is an acute phase serum protein in man, rabbit and most other mammalian species. The phylogenetically conserved, homologous proteins (pentraxins) in all vertebrates studied as well as the invertebrate, Limulus polyphemus, are produced primarily by the liver and share a cyclic pentameric configuration and the property of calcium-dependent binding to phosphate esters such as phosphorylcholine (PC). Species variation is manifest in structural properties such as the tendency of the basic pentameric disc to stack, covalent/non-covalent sub-unit associations, and glycosylation. In those species in which CRP behaves as an acute phase reactant, serum levels increase from 4-1000 fold in response to inflammation and tissue injury (Baltz et al., 1982; Gewurz et al, 1982).

Rat CRP is a constitutive, glycosylated serum protein which shares 70% amino acid sequence homology with human CRP but differs in possessing a single complex oligosaccharide on each monomer. The five monomer subunits are associated by both convalent and non-covalent bonds (Nagpurkar and Mookerjea, 1981; deBeer et al., 1982; Taylor et al., 1984). Rat CRP shows calcium dependent binding to phosphorylcholine (and C-polysaccharide) (de Beer et al., 1981) as well as a calcium dependent association with apoprotein B and E-containing lipoproteins (Rowe et al., 1984; Saxena et al., 1987).

Normal serum levels of 300-600 ug/ml increase 2-3 fold during the acute phase response (deBeer et al., 1982).

METHODS

CRP levels in the serum of Lobund-Wistar rats were determined by standard immunodiffusion (RID) methodology. Rat CRP was purified from pooled normal rat serum (Sprague-Dawley) by a sequence of phosphorylcholine (PC) affinity chromatography and gel filtration as previously described (deBeer et al., 1982; Cabana et al.). Protein was quantitated by Lowry assay and purity confirmed by isoelectric focussing (IEF), SDS-PAGE and by Ouchterlony against monospecific antibodies to rat albumin, transferrin, fibrinogen and C3b. The purified rat CRP was used for immunization of NZW rabbits as described (Nagpurkar and Mookerjea, 1981). For use in the RID assay, the anti-rat CRP plasma pool was absorbed overnight at 4°C with PC-absorbed (4X) normal rat serum. The assay used 1.4% agarose (Seakem ME) in glycine-saline-EDTA buffer (pH 8.2) and a concentration of antibody which produced a linear standard curve between 37-1200 ug/ml, a range encompassing random normal values and experimental acute phase increases. All reported values were calculated from the mean of three determinations.

RESULTS AND DISCUSSION

A summary of the data on CRP levels of Lobund-Wistar rats at varying ages and maintained under germfree or conventional conditions with and without dietary restriction, ie ad libitum feeding or 30% diet reduction, is shown in Table 1.

Table 1. CRP LEVELS AS A FUNCTION OF AGE, ENVIRONMENT AND NUTRITION

ANIMAL GROUP		CRP (ug/ml) 6 MOS	18 MOS	30 MOS	ALL ANIMALS
CONVENTIONAL,	FULL-FED	693±36	705±21	648±33	681
" ",	RESTRICTED	627±21	618±31	599±30	613
GERM-FREE,	FULL-FED	603±12	614±41	521±22	582
" ",	RESTRICTED	601±27	459+28	526±29	526
TOTAL (N = 91)		630	599	575	604

Dietary restriction and a germ free environment significantly reduced serum CRP in healthy rats of all ages, with germfree restricted animals showing the lowest levels. Statistical analysis showed no significant effects of age from 6-30 mos on CRP levels. deBeer et al. (1982) showed no significant differences in CRP levels of pathogen free and conventional rats of several species at the limited ages of 4 and 8 months; greatly reduced levels (<200 ug/ml) were seen in 3 week old animals. The effects of nutritional stress were not tested. The homologous pentraxin, hampster female protein, was detected at 15-20 ug/ml at 5-10 days of age, increased to peak adult levels between 1-1.5 mg/ml at 30-45 days, and then decreased gradually to $\geq$250 ug/ml between 2-12 months (Coe, 1983). The human acute phase CRP response has been reported to be normal with increased age and CRP levels were valuable indicators for infectious disease management in the elderly (Cox et al., 1986).

The presence of CRP in the serum of the normal rat at concentrations higher than acute phase human CRP levels provides a special opportunity to observe both positive and negative metabolic changes. CRP levels, ages, and pathologies of selected moribund animals are detailed in Table 2. Markedly elevated CRP levels (<800 ug/ml) were seen in certain animals with malignancies, particularly of the liver, and significant reductions(<200 ug/ml) were seen in certain animals with liver and hematopoetic abnormalities.

Table 2. RATS WITH MARKED ALTERATIONS OF CRP LEVELS

SERUM CRP*		AGE	DIAGNOSIS
100 ug/ml	-	37 mos	myxoma
140 ug/ml	-	32 mos	prostate adeno-carcinoma
150 ug/ml	-	28 mos	leukemia
170 ug/ml	-	30 mos	hepatoma, pancreatic/tumor
170 ug/ml	-	37 mos	unknown
200 ug/ml	-	41 mos	anemia
850 ug/ml	-	35 mos	hepatoma
870 ug/ml	-	18 mos	normal
940 ug/ml	-	39 mos	pituitary tumor/parathyroid tumor
950 ug/ml	-	38 mos	hepatoma

*Normal = 300-650 ug/ml

The significance of variations in serum CRP levels in this species is not known. Human CRP is an effective precipitin, agglutinin, complement activator, opsonin, and modulator of immune and inflammatory cell function (Gewurz et al., 1982). Rat CRP however does not precipitate or agglutinate PC ligands or mediate complement activation (deBeer et al., 1982). Saxena et al. (1986) reported that the binding of rat CRP to lipoproteins inhibited lipoprotein interactions with liver membrane receptors, suggesting a role for CRP in lipoprotein metabolism. In any case, the rat, with its high constitutive levels of CRP, provides a unique and valuable model for the study of the relationship between nutritional and environmental stress and CRP metabolism.

REFERENCES

Baltz ML, deBeer FC, Feinstein A, Munn EA, Milstein CP, Fletcher TC, March JF, Taylor J, Bruton C, Clamp JR, Davies AJS, Pepys MB (1982). Phylogenetic aspects of C-reactive protein and related proteins. Ann NY Acad Sci 389:49-75.

Cabana VG, Gewurz H, Siegel JN (1982). Interaction of very low density lipoproteins (VLDL) with rabbit C-reactive protein. J Immunol 128: 2343-2348.

Coe JE (1983). Homologs of CRP: A diverse family of proteins with similar structure. Contemporary Topics in Molecular Immunology (ed by Inman FP and Kindt TJ) Plenum Press NY 9:211-238.

Cox ML, Rudd AG, Gallimore R, Hodkinson HM, Pepys MB (1986). Real time management of serum C-reactive protein in the management of infection in the elderly. Age and Aging 16: 257-266.

deBeer FC, Baltz ML, Munn A, Feinstein A, Taylor J, Bruton C, Clamp JR, Pepys MB (1982). Isolation and characterization of C-reactive protein and serum amyloid P component in the rat. Immunology 45: 55-70.

Gewurz H, Mold C, Siegel J, Feidel B (1982). C-reactive protein and the acute phase response. Adv Int Med 27: 345-371.

Nagpurkar A, Mookerjea S (1981). A novel phosphorylcholine binding proteins from rat serum and its effects on heparin-lipoprotein complex formation in the presence of calcium. J Biol Chem 256: 7440-7445.

Rowe IF, Soutar AK, Trayner IM, Baltz ML, deBeer FC, Walker L, Bowyer D, Herbert J, Feinstein A, Pepys MB (1984). Rabbit and rat C-reactive proteins bind apolipoprotein B-containing lipoproteins. J Exp Med 159: 604-616.

Saxena U, Nagpurkar A, Mookerjea S (1986). Inhibition of the binding of low density lipoproteins to liver membrane receptors by rat serum phosphorylcholine binding protein. Biochem Biophys Res Commun 141: 151-157.

Taylor JA, Bruton CJ, Anderson JK, Mole JE, deBeer FC, Baltz ML, Pepys MB (1984). Amino acid sequence homology between rat and human C-reactive protein. Biochem J 221: 903-906.

ENDOCRINOLOGY

THE EFFECT OF DIETARY RESTRICTION ON SERUM HORMONE AND BLOOD CHEMISTRY CHANGES IN AGING LOBUND-WISTAR RATS

David L. Snyder and Beth Towne.

Lobund Laboratory, University of Notre Dame, Notre Dame IN 46556.

Serum hormone and blood chemistry parameters can provide valuable information on the physiological status of aging animals. They can be used as innocuous tests for detecting individual differences in biological aging rate (Short et al., 1987), and as indicators of kidney pathology (Kafetz, 1984), nutritional status (Kergoat et al., 1987), and endocrine function (Sonntag, 1987). For these reasons serum hormone and blood chemistry parameters were determined in our study of the effects of mild dietary restriction (DR) on aging in germfree (GF) and conventional (CV) male Lobund-Wistar (L-W) rats. The analysis of serum hormones is particularly critical since it has been proposed that food restriction is coupled to the aging process through changes in the endocrine and neuroendocrine regulatory systems (Masoro, 1988).

Serum may provide useful biological markers of the aging process, but these markers do not reveal the mechanisms behind the aging process. Many of these markers are the result of biochemical alterations and disease processes which may only be partially related to aging. The data reported here will aid in the interpretation of the results of the other examinations being performed on tissues distributed through the Lobund Aging Study. Hopefully these findings will lead to new experiments designed to uncover the specific mechanisms associated with DR, aging and the endocrine system and eventually to methods of intervention in the aging process.

In this paper we report on the serum levels of insulin (Ins), thyroxine (T4), triiodothyronine (T3), testosterone (T), thyroid stimulating hormone (TSH), prolactin (PRL) and automated blood chemistry profiles in fasted male L-W rats between 7 and 27 months of age.

METHODS

Details on the housing, feeding and sacrifice of the L-W rats used in this study are reported in these proceedings in the chapter by Snyder and Wostmann. F refers to full-fed (ad libitum intake) rats and R refers to the restricted fed rats (70% of adult intake). During the first three years of the study, several GF isolators became contaminated with single strains of microorganisms. *Actinomyces sp.* or a gram positive coccus were cultured in these isolators. Previous experience suggests that these microorganisms were not numerous enough to alter the normal GF characteristics. Rats contaminated with these microorganisms were immediately sacrificed and used for preliminary studies of hormone and blood chemistry parameters. The contaminated rats will be referred to as gnotobiotic (GN) instead of strictly germfree. Comparable CV rats were sacrificed or orbital blood was obtained for comparison.

A fresh portion of serum was assayed for blood chemistry parameters by a computerized sequential multiple analyzer at the South Bend Medical Foundation. The following values were determined and used for analysis: glucose, uric acid, blood urea nitrogen, creatinine, sodium, potassium, chloride, CO_2 content, inorganic phosphorous, calcium, triglycerides, cholesterol, total bilirubin, alkaline phosphatase, gamma-glutamyl transferase, aspartate transaminase, alanine transferase, total protein, albumin, and globulin. Serum for hormone analysis was frozen at $-70^{\circ}C$ and stored for several months prior to use.

Commerical antibody coated-tube radioimmunoassay kits intended for clinical use were used to determine serum T4, T3, T and Ins. These assays offer the convenience of short incubation time, provide low nonspecific binding, eliminate second antibodies or other separation procedures and do not require centrifugation. There is negligible crossreactivity with other hormones when using these kits. Kits were purchased from Diagnostic Products Corporation (DPC, Los Angeles, CA).

The insulin assay kit was modified for use with rat serum. Purified rat insulin from Novo Biolabs Division of Novo Laboratories, Inc. (Wilton, CT) was used to prepare the assay standards. The insulin had an activity of 24.3 IU per mg and had been purified on a Sephadex column to remove proinsulin and related substances. Insulin standards of 6, 4, 3, 2, 1 and 0.4 ng/ml were prepared in 0.04M phosphate buffered saline with 6% bovine serum albumin. This buffer was also used to determine maximum binding. Standards were always prepared fresh the day before each assay was performed. Since rat serum has an enhanced ability to destroy the ^{125}I-insulin tracer the assay had to be prepared on ice and incubated at 4°C. After several trial runs, 18 hours (overnight) was deemed the best incubation time to reach maximum binding.

A number of parameters for the insulin assay were determined. For all assays combined the maximum binding was 26%, the intraassy coefficient of variation was 8.8% and the interassay coefficient of variation 12.3%. The coefficients of variation of serum spiked with none, 2 ng/ml and 6 ng/ml of rat insulin were 7.4%, 12.4% and 8.0% respectively. The least detectable dose estimated as two standard deviations from maximum binding was 5.7 uIU/ml, while the least detectable dose estimated as 95% of maximum binding was 3.7 uIU/ml. The least detectable dose was based on the mean maximum binding of 20 zero standard tubes run in the same assay. Rat serum spiked with 6, 3, 2, 1, 0.5 and 0.2 ng/ml insulin run in two assays yielded slopes of 1.03 and 1.56, intercepts of 5.9 and 6.2 uIU/ml and correlation coefficients of .99 when the spiked concentration was plotted against the concentration estimated by the assay.

The Ins levels as determined by the DPC kit are similar to values reported for the L-W rat using a double antibody kit and the NOVO rat insulin standard (Sewell, 1976). In this study values for Ins from 3 month old rats were 55.0±8.8 uU/ml with accompanying glucose of 122±4 mg/100ml. The DPC Coat-A-Count Insulin Assay Kit gives accurate results to 10 uIU/ml when used at refrigerator temperatures with a rat insulin standard. When used for determining human insulin levels this kit is accurate to 3 uIU/ml which is similar to the value obtained in our assay. However our calculations are based on a maximum binding in aqueous buffer and not rat serum. Rat serum spiked with 5 uIU/ml of insulin was detectably different from rat serum which had not been spiked. The best kits were those obtained shortly after iodination and have the highest maximum counts.

Serum PRL and TSH were determined by double-antibody radioimmunoassay with materials from kits provided by the National Hormone and Pituitary Program (NIADDK). Reference preparation NIADDK-rPRL-RP-3 was used in the PRL assay and is equal in potency to rPRL-RP-2 and 2.8 times more potent than rPRL-RP-1. Reference preparation NIADDK-rTSH-RP-2 was used in the TSH assay and is 176 times more potent than rTSH-RP-1.

Data for the blood chemistry parameters and serum Ins were analyzed using the general linear models program from the Statistical Analysis System software package (SAS Institute, 1985). This program performs analysis of variance for unequal sample sizes. A multifactorial design with three main effects (age, diet, and microbial status) was used to determine if the treatments within each main effect had a significant effect on the parameter being analyzed. Age had three treatments (adult: 7 to 13 months; middle: 16-20 months; old: 24-27 months), diet had two treatments (F and R) and microbial status had two treatments (GN vs CV). Duncan's multiple range test was used to distinguish between treatments when factors were deemed significant in the analysis of variance. Significance was set at P<.01. Table 1 gives the number of rats, average age and average body weight in each of the groups used in the analysis of the blood chemistry parameters and serum Ins. Means and standard errors were calculated from serum hormone data organized according to the age groups mentioned above. No statistical analysis was performed on the hormone data because they were compiled from a number of discrete experiments which will be published later.

TABLE 1. Characteristics of the male Lobund-Wistar rats used to determine blood chemistry parameters and serum insulin: number of rats, average age in months, and body weight in grams.

	Conventional Full-fed	Conventional Restricted	Gnotobiotic Full-fed	Gnotobiotic Restricted
Adult (7-13 months)				
N	28	25	43	19
Age	9.9	10.1	9.5	9.3
Weight	436	286	394	278
Middle (14-23 months)				
N	10	15	8	16
Age	19.3	18.0	18.0	19.0
Weight	484	295	442	328
Old (24-27 months)				
N	12	20	10	10
Age	24.0	24.5	24.8	26.6
Weight	469	289	479	333

RESULTS

Table 2 contains the blood chemistry parameters which were significantly altered either by microbial status, DR or age. Serum glucose rose with age in the F rats but not the R rats for GN and CV data combined. There were no differences in serum glucose levels between GN and CV rats.

Serum triglycerides and cholesterol rose with age in F rats only, and were higher in CV rats compared to GN rats and in F rats compared to R rats. Inorganic phosphorous was higher in GN rats and decreased with age in all groups. Alkaline phosphatase increased with age in all groups and was higher in CV and R groups. The age-related changes in inorganic phosphorous and alkaline phophatase may reflect a decrease in the rate of bone formation with age (Hodkinson, 1984) that is not affected by DR (see the chapter by Nishimoto in these proceedings). Total serum protein and globulin were elevated in CV and F groups. Since serum albumin was not altered by DR the changes in total serum protein are due to changes in serum globulin. The reduction in serum globulin in GN and in R rats may be due to a lack of antigenic stimulus especially in the GN rats.

TABLE 2. Blood chemistry parameters (mean±standard error) of male L-W rats which were significantly altered by microbial status, dietary restriction, or age. Statistically significant effects are given after each parameter.

Parameter/ Age group	CV-F	CV-R	GN-F	GN-R
Glucose (mg/100ml):				
Adult	131±2	136±2	131±3	121±5
Middle	160±13	145±5	165±3	163±6
Old	163±6	133±3	159±5	131±10

Statistics: middle and old full-fed > adult full-fed.

Parameter/ Age group	CV-F	CV-R	GN-F	GN-R
Triglycerides (mg/100ml):				
Adult	126±6	101±5	92±4	73±5
Middle	118±10	95±6	142±14	81±9
Old	167±17	94±5	115±12	77±11

Statistics: CV>GN; F>R; Old CV-F > adult CV-F and middle CV-F.

Parameter/ Age group	CV-F	CV-R	GN-F	GN-R
Cholesterol (mg/100ml):				
Adult	94±2	98±2	85±2	78±2
Middle	116±4	95±2	105±3	98±4
Old	142±7	114±2	115±4	86±5

Statistics: CV>GN; F>R; Old full-fed>middle full-fed>adult full-fed.

Table 2 continued.

Parameter/Age group	CV-F	CV-R	GN-F	GN-R
Inorganic phosphorous (mg/100ml):				
Adult	5.8±0.2	5.4±0.1	6.7±0.1	7.5±0.2
Middle	5.6±0.2	5.7±0.2	---	6.2±0.1
Old	5.1±0.1	5.2±0.1	5.7±0.1	6.3±0.4

Statistics: GN>CV; Adult>middle>old for all groups combined.

Alkaline phosphatase (IU/L):				
Adult	204±15	230±12	101±3	130±1
Middle	196±3	249±21	146±10	190±15
Old	231±1	298±8	156±9	129±13

Statistics: CV>GN; R>F; Old>middle>adult for all groups combined.

Total protein (g/100ml):				
Adult	6.2±0.1	6.1±0.1	5.7±0.1	5.5±0.1
Middle	6.2±0.1	5.7±0.2	5.8±0.1	5.6±0.1
Old	6.3±0.1	6.1±0.0	5.8±0.1	5.6±0.1

Statistics: CV>GN; F>R.

Globulin (g/100ml):				
Adult	2.6±0.0	2.3±0.0	1.9±0.0	1.7±0.0
Middle	2.4±0.1	2.2±0.0	1.8±0.0	1.9±0.1
Old	2.7±0.1	2.4±0.0	2.1±0.0	1.9±0.0

Statistics: CV>GN; F>R.

Table 3 lists the blood chemistry parameters which were significantly affected by gnotobiotic status but showed no changes due to DR or age. Uric acid, potassium, chloride and albumin were higher in GN rats, while CO_2 content was lower in GN rats. These data reflect the changes in ionic balance and the slight hemoconcentration within the blood of GF rats that compensate for the loss of water to the enlarged cecum (Wostmann, 1975). Table 4 lists the blood chemistry parameters which showed no changes due to microbial status, DR or age. As a whole the parameters listed in Table 4 show that kidney and liver function in the L-W rats were stable up to 27 months.

TABLE 3. Blood chemistry parameters of male Lobund-Wistar rats which were significantly altered by microbial status only (combined mean).

Parameter	Conventional	Gnotobiotic
Uric acid (mg/100ml)	1.4	1.8
Potassium (mEQ/L)	5.5	6.1
Chloride (mEQ/L)	107	109
Albumin (g/100ml)	3.7	3.8
CO_2 content (mEQ/L)	24.1	22.2

TABLE 4. Blood chemistry parameters of male Lobund-Wistar rats which were unaltered by microbial status, dietary restriction or age.

Parameter	Range *
Calcium (mg/100ml)	8.6 - 11.0
Sodium (mEQ/L)	141 - 148
Bilirubin (mg/100ml)	0.8 - 1.2
Blood urea nitrogen	12.6 - 15.4
Creatinine (mg/100ml)	.35 - .53
Gamma-glutamyl transferase (IU/ml)	0.4 - 3.2
Aspartate transaminase (IU/ml)	106 - 150
Alanine transferase (IU/ml)	40 - 70

*lowest and highest individual group means

Serum Ins levels corresponding to the rats used in the blood chemistry analysis are listed in Table 5. The only statistically significant effect on serum Ins was an overall increase in F rats compared to R rats.

TABLE 5. Serum insulin levels in male Lobund-Wistar rats (uIU/ml; mean±standard error).

Age Group	CV-F	CV-R	GN-F	GN-R
Adult	35.4±1.1	34.1±1.4	34.5±2.4	31.8±2.2
Middle	38.6±3.1	25.3±2.3	40.5±1.6	30.4±2.9
Old	34.1±3.5	32.3±2.9	27.6±2.2	26.6±3.1

Statistics: F > R.

The composite serum hormone data are presented in Table 6. Serum T4 declined with age in all but the GF-R rats, was slightly higher in GF rats and was not affected by DR. Serum T3 was unaffected by age and DR within the CV rats. However the GF rats when compared to the CV rats had higher serum T3 levels in the adult age categories. These levels dropped to similar CV levels by middle age. Serum TSH was generally unaffected by age, DR or microbial status. Serum PRL rose with age in F rats but not in R rats and was not affected by microbial status. Serum T showed a consistent fall with age in all four groups, was higher in R rats at all ages and was slightly higher in GF rats.

TABLE 6. Serum hormones levels (mean±standard error) in male Lobund-Wistar rats. Sample size from 7 to 40 rats.

Hormone/ Age Group	CV-F	CV-R	GF-F	GF-R
Thyroxine (ug/100ml)				
Adult	4.5±0.1	4.4±0.1	5.0±0.2	5.3±0.2
Middle	4.1±0.2	3.5±0.1	4.5±0.2	4.5±0.3
Old	3.4±0.2	3.7±0.1	3.8±0.2	4.8±0.1
Triiodothyronine (ng/100ml)				
Adult	86±5	99±3	112±5	124±5
Middle	99±7	96±4	89±5	107±10
Old	94±9	106±5	89±7	91±3
Thyroid Stimulating Hormone (ng/ml)				
Adult	2.1±0.1	2.1±0.2	2.4±0.2	2.4±0.2
Middle	1.7±0.1	2.0±0.2	2.6±0.3	2.2±0.1
Old	1.7±0.2	1.7±0.4	1.6±0.3	1.8±0.3
Prolactin (ng/ml)				
Adult	29±3	25±5	22±1	25±2
Middle	43±6	21±3	31±4	42±2
Old	52±11	28±2	73±21	31±6
Testosterone (ng/ml)				
Adult	3.2±0.4	5.4±0.6	3.5±0.2	7.4±0.8
Middle	3.2±0.4	4.3±0.9	1.8±0.2	5.6±1.2
Old	1.5±0.3	2.9±0.4	1.4±0.2	2.6±0.2

DISCUSSION

The data presented here reflect the condition of healthy male L-W rats up to 27 months of age. These rats showed no evidence of prostatitis, prostate tumors or mammary adenofibromas, conditions which are first observed in L-W rats after 18 months of age. Nephrosis is infrequent in L-W rats and was not present in the rats used in the present study. Therefore changes in blood chemistry and serum hormones are more reflective of age-related changes than of disease-related changes. Although these changes do not directly describe the mechanism of DR's effect on aging, they do give several indications where DR works.

The blood chemistry analysis revealed that age-related increases in serum glucose, triglycerides and cholesterol were prevented by a reduction in food intake of only 30%. These findings are similar to other reports of diet-restricted Fischer 344 rats (Liepa et al., 1980; Masoro et al., 1983) and Sprague-Dawley rats (Reaven and Reaven, 1981). Changes in serum glucose and lipids may be due to changes in circulating hormone levels. Reaven and Reaven (1981) have suggested that DR enhances insulin sensitivity and therefore prevents the age-related rise in hepatic triglyceride secretion. Our study also provides evidence for this assumption since Ins levels were reduced by DR while serum triglycerides and glucose remained stable into old age. Hypothyroidism is an important cause of secondary hyperlipidemia in the elderly especially if kidney and liver disease are not present (Thompson, 1984). Hypothyroidism may be a cause of increased serum triglycerides and cholesterol in old CV-F L-W rats since serum T4 levels are the lowest in this group and no liver or kidney disease was present. The data on serum T3, the active thyroid hormone, indicate however that age and DR had little effect on this hormone. More study is needed to determine the extent of thyroid function changes in aging L-W rats and the effect of DR on these changes.

Endocrine and neuroendocrine mechanisms have been suggested as the coupling agent between DR and extended life span (Masoro, 1988). The absence of reduced serum TSH, T4, T3 and T levels in diet-restricted L-W rats argues against the theory that DR extends life span by reducing pituitary hormone secretion. In our study DR enhanced T levels and maintained normal TSH and T3 levels. Testes size was not affected by DR indicating that serum gonadotropin levels were enhanced or the testes were more sensitive to gonadotropins in DR rats. The ability of DR to prevent the age-related rise in serum PRL supports the connection between DR and the neuroendocrine system. The low numbers of PRL adenomas found in L-W rats older than 30 months (see the chapter by Kovacs in these proceedings) suggest that the rise in serum PRL with age is related to hypothalamic regulation of PRL secretion and not the development of pituitary adenomas. Therefore DR may act by preventing age-related changes or damage to the neuroendocrine mechanisms responsible for regulating pituitary function.

The data presented here provide an initial understanding of changes in endocrine function during aging and DR in the L-W rat. The conventional L-W rat provides an excellent model for examining the interaction between nutrition, endocrine function and the aging process. Further study is needed to examine alterations in the biologic activity of hormones, post-translational processing of hormones, molecular mechanisms of protein synthesis, and intracellular mechanisms of hormone action in ad libitum and diet-restricted rats. These types of studies may eventually lead to an understanding of the mechanisms for DR's ability to extend life span.

REFERENCES

Hodkinson M (1984). Calcium, phosphate and the investigation of metabolic bone disease. In Hodkinson M (ed.): "Clinical Biochemistry of the Elderly," Edinburgh:Churchill Livingstone, pp 139-152.

Kafetz K (1984). Electrolytes, urea, creatinine and uric acid. In Hodkinson M (ed.): "Clinical Biochemistry of the Elderly," Edinburgh:Churchill Livingstone, pp 167-194.

Kergoat M, Leclerc BS, PetitClerc C, Imbach A (1987). Discriminant biochemical markers for evaluating the nutritional status of elderly patients in long-term care. Am J Clin Nutr 46:849-61.

Liepa GU, Masoro EJ, Bertrand HA, Yu BP (1980). Food restriciton as a modulator of age-related changes in serum lipids. Am J Physiol 238 (Endocrinol Metab 1):E253-257.

Masoro EJ (1988). Food restriction in rodents: an evaluation of its role in the study of aging. J Gerontol: Biol Sci 43:B59-B64.

Masoro EJ, Compton C, Yu BP, Bertrand H (1983). Temporal and compositional dietary restrictions modulate age-related changes in serum lipids. J Nutr 113:880-892.

Reaven GM, Reaven EP (1981). Prevention of age-related hypertriglyceridemia by caloric restriction and exercise training in the rat. Metabolism 30:982-986.

SAS Institute Inc. (1985). "SAS User's Guide: Statistics, Version 5 Edition." Cary, NC: SAS Institute, Inc., pp 433-506.

Sewell DL, Bruckner-Kardoss E, Lorenz LM, Wostmann BS (1976). Glucose tolerance, insulin and catecholamine levels in germfree rats. Proc Soc Exp Biol Med 152:16-19.

Short R, Williams DD, Bowden DM (1987). Cross-sectional evaluation of potential biological markers of aging in pigtailed macaques: effects of age, sex and diet. J Gerontol 42:644-654.

Sonntag WE (1987). Hormone secretion and action in aging animals and man. Rev Biol Res Aging 3:299-335.

Thompson GR (1984). Cholesterol and the plasma lipoproteins. In Hodkinson M (ed.): "Clinical Biochemistry of the Elderly," Edinburgh:Churchill Livingstone, pp 296-311.

Wostmann BS (1975). Nutrition and metabolism of the germfree mammal. World Rev Nutr Diet 22:40-92.

AGE-RELATED CHANGES IN ADRENAL CATECHOLAMINE LEVELS AND MEDULLARY STRUCTURE IN MALE LOBUND-WISTAR RATS

Nancy P. Nekvasil and Toni R. Kingsley

Biology Department, St. Mary's College, and South Bend Center for Medical Education, University of Notre Dame, Notre Dame, Indiana 46556

INTRODUCTION

Aging is associated with a gradual decrease in the ability of various tissues and organ systems to respond to stressful situations (Finch, 1976; Selye and Tuchweber, 1976). This progressive decline in physiological processes includes endocrine systems, such as the adrenomedullary-sympathetic system. These changes include an increase in both epinephrine (E) and norepinephrine (NE) in adrenal glands from aged male F344 rats (Martinez et al., 1981) and an age-related increase in dopamine-β-hydroxylase (DBH) activity (an enzyme which converts dopamine to NE) in three species of rodents (Banerji et al., 1984). Thus, it is apparent that catecholamines (CA) are somehow involved in the aging process though the exact relationship is unknown.

The development of adrenomedullary tumors and hyperplasia with age has been reported for rats of varying strains. A characteristic pattern of benign tumors in endocrine organs of male Lobund-Wistar rats raised in a germ-free environment has emerged (Pollard and Wostmann, 1985). More than 90% of those animals at 36 months of age exhibit tumors or enlargement of the adrenal medulla. Tischler et al. (1985) report that aging Long-Evans rats develop hyperplasia of the adrenal medulla. The altered gland is characterized by sparse E content and small granules in adrenomedullary cells. In an EM study, Shaposhnikov (1985) reports no significant age-related differences in the ultrastructure of chromaffin cells in the adrenal medulla of Wistar rats. Thus, while it is agreed that age-related alterations in gland size occur in most laboratory rats, a concurrent ultrastructural alteration is inconclusive.

It might be expected that with the appearance of an enlarged adrenal gland, CA synthesis would be altered in some fashion. Even with the increase in gland size, Bosland and Bar (1984) report that the aged Wistar rats with adrenomedullary tumors show no significantly altered endocrine activity. They also conclude that excessive CA synthesis and release are not features of the enlarged gland which occurs spontaneously as aging progresses. However, Banerji et al. (1984) reported elevated DBH activity in adrenal glands of aged rats, hamsters, and gerbils. At best, the information available is scarce and contradictory as to the functional activity of adrenal glands from aging animals.

This study was designed to examine the changes in adrenal gland size and CA content relative to aging in Lobund-Wistar rats. These animals were raised in different environmental conditions (germfree or conventional status) under varying feeding regimens (fullfed or restricted diet) to determine a more complete relationship between aging, diet, and the adrenomedullary changes that have been noted.

MATERIALS AND METHODS

Male Lobund-Wistar rats were raised in either germfree (G) or conventional (C) environments and fed <u>ad libitum</u> (F) or a 30% restricted diet (R). Restricted animals were housed separately to ensure uniform feeding, and all animals were exposed to a 12:12 light:dark cycle. Animals were raised to 6, 18, or 30 months of age and were then sacrificed for examination of adrenal CA levels. All animals were sacrificed between 9 and 10am.

The rats were weighed and then decapitated immediately to eliminate undo handling and excess stress. Adrenal glands were removed immediately, designated as right or left, cleanly dissected, put in preweighed microhomogenizer tubes, and placed on dry ice. After weighing the tubes, 0.1ml sodium metabisulfite (4μmol) was added to prevent CA oxidation, and the glands were homogenized. Deproteinization of the sample was accomplished through the addition of 0.8ml perchloric acid (0.1M), and 0.1ml dihydroxybenzylamine (DHBA; 10μg) was then added as an internal standard to a total volume of 1ml. Homogenized samples were transferred to cold Sorvall tubes and placed on ice until all samples were prepared. Centrifugation at 0°C and 30,000xg followed for 20 minutes to produce a clear supernatant. The supernatant was aspirated off the precipitate, filtered through a 45μm HPLC syringe-fit filter (Waters) and deposited into a 2ml plastic screw-top vial. All samples were placed in a -80°C freezer from 2 days to 2 weeks until they could be assayed on the HPLC.

Analysis of the samples was accomplished by injecting 0.1ml of a diluted (1:50) portion and an undiluted portion into an HPLC system. The HPLC (Waters 840) was equipped with a C18 NovaPak column and an ESA Coulachem electrochemical detector (Model 5100A). The diluted portion was used to separate and quantitate dihydroxymandelic acid (DHMA), NE, E, and DHBA. Dopamine (DA) was quantitated from the undiluted portion.

The mobile phase used to elute the CA and their metabolites consisted of 6.9g sodium phosphate monobasic, 20mg EDTA, 1.5g heptanesulfonic acid, and 70ml HPLC grade methanol (for a 7% methanol mobile phase) diluted to 1 liter. The pH of the mobile phase was adjusted to 3.6 with 3N HCl. The mobile phase was filtered through a 45μm HPLC grade filter and degassed prior to use.

RESULTS

Statistical analysis revealed that every adrenal parameter measured showed significant variation ($p=0.01$ or $p=0.05$) when the data were grouped according to age (Table 1). Catecholamine content is defined as the total amount of CA (μg) in the entire adrenal gland. Catecholamine concentration is the amount of CA per milligram of adrenal tissue (ng/mg). Norepinephrine content, NE concentration, and E content were found to be significantly higher ($p=0.05$, $p=0.01$, $p=0.01$ respectively) in the C animals than the G rats when grouped by status (C vs.G). Only one parameter demonstrated significant variation when compared by diet (F vs. R). Epinephrine content is significantly greater ($p=0.01$) in the F rats.

Significant enlargement of adrenal glands was observed only for the 30 month (m) animals (Fig. 1). No younger animals had enlarged adrenals.

Table 1. Statistical analysis of adrenal parameters comparing age, status, and diet in Lobund-Wistar rats.

STATISTICAL ANALYSIS

PARAMETER	FACTOR	ANOVA*	DUNCAN**
adrenal weight	age	++	30>6; 6=18
dopamine content	age	++	30>18; 18=6
dopamine concentration	age	++	30>18; 18=6
norepi. content	age	++	30>6; 6=18
	status	+	C>G
norepi. concentration	age	++	30>6; 6=18
	status	++	C>G
epi. content	age	++	30>18>6
	status	++	C>G
	diet	++	F>R
epi. concentration	age	+	18>30; 6=18&30
epi/norepi content	age	++	18>6; 6=30
epi/norepi concentration	age	++	18>6>30

*Multifactorial analysis of variance (SAS Institute Inc., 1985): $P<0.01$ = ++, $P<0.05$ = +.
** Duncan mean separation within significant factor: $P<0.05$

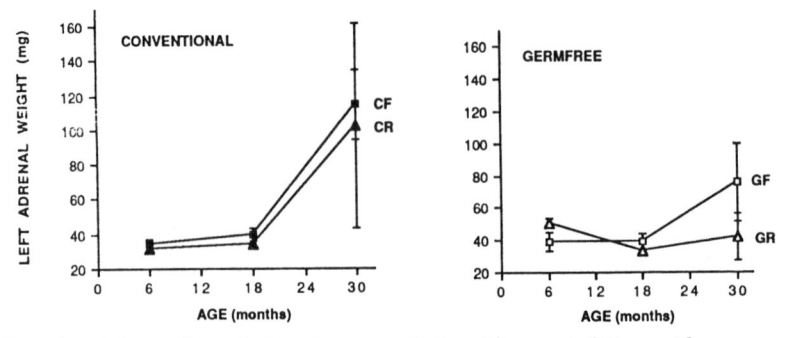

Fig. 1. Adrenal weight in mg of 6, 18, and 30 month conventional and germfree Lobund-Wistar rats.

Catecholamine profiles differ considerably between the 6m, average-sized (31.8 ± 2.7 mg) glands and the 30m, enlarged (102.4 ± 59.3 mg) adrenal glands (Table 2). Adrenal weight in 30m animals shows a 3-fold increase over the 6m rats. Norepinephrine content increases by 8X and DA increases 20-fold in the 30m animals. Epinephrine content remains nearly the same even though its precursors are elevated in the older rats. The metabolite, DHMA, decreases slightly in older animals, and the E/NE ratio also declines from 2 in the 6m group to 1 in the 30m glands that display hypertrophy.

Table 2. Catecholamine content in average, normal-sized adrenal glands (6m) compared to adrenals displaying hypertrophy (30m).

	6 MOS.	30 MOS.	% CHANGE
L. ADR. WT. (mg)	31.8 ± 2.7	102.4 ± 59.3	322%
EPI CONTENT (µg)	15.0 ± 0.9	16.8 ± 2.6	112%
NOREPI CONTENT (µg)	7.2 ± 1.5	55.0 ± 31.9	760%
DA CONTENT (µg)	0.09 ± 0.01	1.7 ± 1.3	1880%
DHMA CONTENT (µg)	37.9 ± 3.0	24.2 ± 1.4	64%
EPI/NOREPI	2.2 ± 0.3	1.0 ± 0.7	45%

*Representative group selected = conventional restricted (n=4). Values are mean ± standard error.

Norepinephrine and DA concentrations are highest in the 30m animals (Figs. 2 and 3). Epinephrine concentration overall decreases in the 30m, CF, CR, and GR animals. However, in the GF group, the 18m rats show a decrease while the 30m rats demonstrate an increase in E concentration. (Fig. 4).

Not all 30m rats demonstrate enlargement of the adrenal glands. Regrouping of the data from the 30m rats into small vs. enlarged glands reveals that the pattern described for CA concentration profiles in the aged rats is more prominent in the enlarged adrenals (Fig. 5).

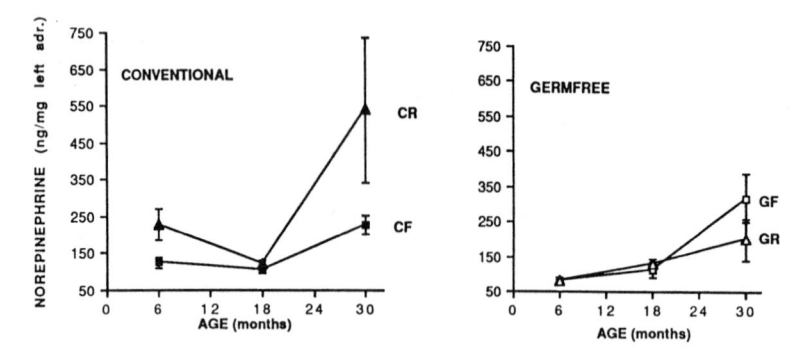

Fig. 2. Adrenal norepinephrine concentration in 6, 18, and 30m conventional and germfree Lobund-Wistar rats.

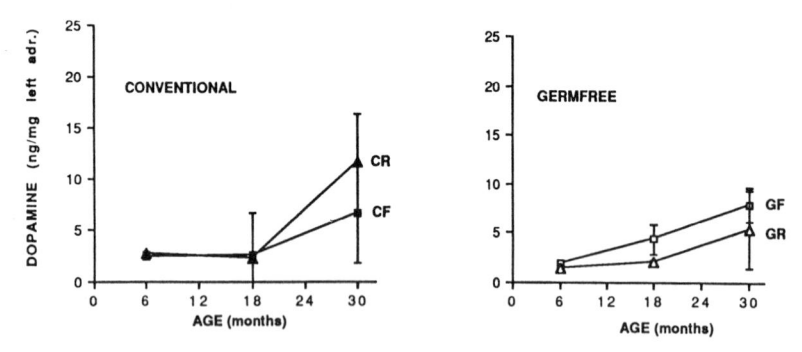

Fig. 3. Adrenal dopamine concentration in 6, 18, and 30m conventional and germfree Lobund-Wistar rats.

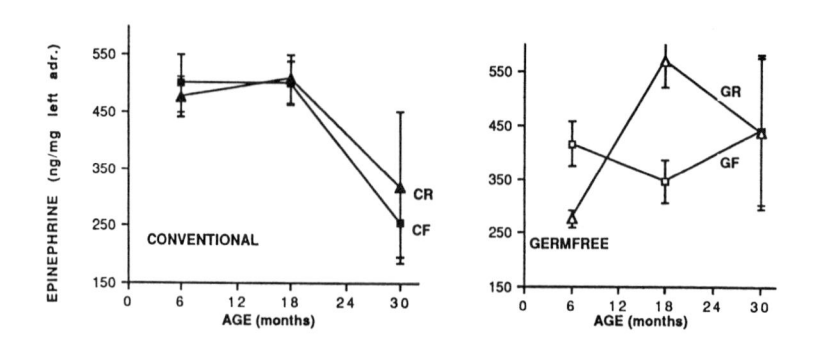

Fig. 4. Adrenal epinephrine concentration in 6, 18, and 30m conventional and germfree Lobund-Wistar rats.

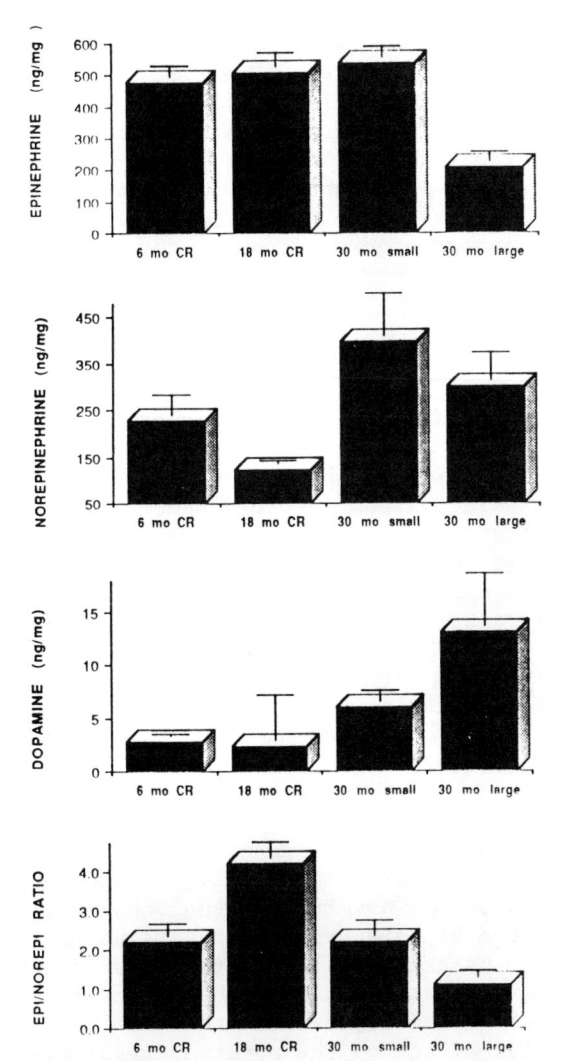

Fig. 5. Catecholamine concentration comparisons of 6, 18, and 30m small, and 30m enlarged adrenal glands in a representative (CR) group.

In the CR group (Fig. 5), E concentrations in the 30m small gland are very similar to the 6 and 18m animals whereas the 30m enlarged gland has much less E per mg of tissue. Dopamine concentrations are elevated in the larger glands whereas NE levels are greater in the small and less in the enlarged 30m glands. Overall, NE levels at 30m are greater than those at either 6 or 18m. The E/NE ratio is greatest for the 18m adrenals and least in the 30m, large glands. The 6m and 30m, small glands have similar values. Of the 32 glands removed from 30m animals, 17 weighed less than 60g (designated as small) and 14 weighed more than 60g (classified as large). One of the 32 glands was an overt tumor and was not grouped with any of the other adrenals.

DISCUSSION

The changes in adrenomedullary function which have been demonstrated in this investigation appear to be inherent to the aging process. All adrenal parameters tested show a significant alteration in the 30m animals suggesting a provision for handling stress into old age. Aging has been historically associated with a decreased ability to deal with stressful situations effectively. It appears that the adrenal hyperplasia found only in the 30m animals is a mechanism for maintaining adequate levels of E in the face of decreasing E synthesizing capacity. Both precursors of E (NE and DA) are elevated in the 30m adrenal glands. This suggests that in order to keep E content at a functional level the precursor molecules must increase in quantity to make up for a decrease in enzyme-converting activity between NE and E.

There are only three parameters demonstrating significance when grouped by status and only one when examining dietary effects on the observed adrenal changes. This suggests that the differences in the various rat groups are little modified by the germfree status, and that the conventional rats may be better able to cope with stress as aging progresses. Additionally, dietary restriction appears to have no positive effect on the observed results. The one parameter (E content) demonstrating significant variation with diet was greater for the fullfed animals than for the restricted group. Apparently in this one aspect of aging, neither a germfree environment nor a restricted diet aid in the aging process to any great degree.

Not all 30m rats develop adrenal hyperplasia. This made it possible to study the differences between aged glands which were normal in size and aged adrenals which were enlarged. The CA profiles are quite distinct, and, in fact, the 30m, small glands are most similar in their CA concentrations to those of the younger adrenal glands. The aging pattern which emerged with this study is most prominent in the 30m enlarged adrenal glands. Epinephrine concentration is only decreased substantially for the enlarged 30m glands while DA is increased. Curiously, NE concentration is elevated for the 30m, small adrenals as well as for the larger glands. It may be possible that the enlarged adrenal occurs in older animals both as a compensatory hypertrophy for maintaining E stores as needed and for keeping NE amounts up when the ability to produce E fails. Maybe the development of the hyperplasia is actually instigated by the alteration in the E/NE ratio. As either of these two biogenic amines change significantly, perhaps the gland responds by producing more cells which in turn produce more NE.

In conclusion, age has the greatest influence on adrenomedullary size and CA content differences. Germfree environment and dietary restriction have little influence on the observed CA variations. The 30m animals are the only ones to develop adrenal hyperplasia. Adrenal enlargement, which occurred in about 44% of the 30m animals, may occur in response to a change in the E/NE ratio. This change could be a result of a decrease in E synthesizing capacity. Both DA and NE levels are increased in 30m animals while E content does not change and E concentration decreases compared to younger animals. Small adrenal glands from 30m animals display a CA profile more similar to 6m and 18m adrenals while the enlarged 30m glands demonstrate an aging pattern with altered amine distribution.

REFERENCES

Banerji T, Parkening T, Collins T (1984). Adrenomedullary catecholaminergic activity increases with age in male laboratory rodents. J Geron 39:264-268.

Bosland M, Bar A (1984). Some functional characteristics of adrenal medullary tumors in aged male Wistar rats. Vet Pathol 21:129-140.

Finch C (1976). The regulation of physiological changes during mammalian aging. Quar Rev of Biol 51:49-83.

Martinez J, Vasquez B, Messing R, Jensen R, Liang K, McGaugh J (1981). Age-related changes in the catecholamine content of peripheral organs in male and female F344 rats. J Geron 36:280-284.

Pollard M, Wostmann BS (1985). Aging in germfree rats: the relationship to the environment, diseases of endogenous origin, and to dietary modification. In Archibald J, Ditchfield J, Rowsell HC (eds): "The Contribution of Laboratory Animal Science to the Welfare of Man and Animals," New York: Gustav Fischer Verlag, pp 181-186.

SAS Institute Inc. (1985). "SAS User's Guide: Statistics, Version 5 Edition." Cary, NC: SAS Institute Inc., pp 529-557.

Seyle H, Tuchweber B (1976). Stress in relation to aging. In: Everitt AV, Burgess JA (eds): "Hypothalamus, Pituitary, and Aging," Springfield, Illinois: Charles C. Thomas.

Shaposhnikov V (1985). The ultrastructural features of secretory cells of some endocrine glands in aging. Mech Ageing and Devel 30:123-142.

Tischler A, DeLellis R, Perlman R, Allen J, Costopoulos D, Lee Y, Nunnemacher G, Wolfe H, Bloom S (1985). Spontaneous proliferative lesions of the adrenal medulla in aging Long-Evans rats. Lab Invest 53:486-498.

MODEST DIETARY RESTRICTION AND SERUM SOMATOMEDIN-C/INSULIN LIKE GROWTH FACTOR-I IN YOUNG, MATURE AND OLD RATS

T. Elaine Prewitt[1] and A. Joseph D'Ercole[2]

[1]Department of Nutrition and Medical Dietetics, College of Associated Health Professions, University of Illinois at Chicago, Chicago, IL 60680
[2]Department of Pediatrics, School of Medicine, University of North Carolina, Chapel Hill, NC 27514

INTRODUCTION

Several lines of evidence have demonstrated that suboptimum dietary intake in experimental animals and humans significantly reduces serum somatomedin-C/Insulin Like Growth Factor 1, (SMC/IGF 1), a growth hormone dependent anabolic peptide (Grant et al., 1973; Phillips et al., 1979; Prewitt et al., 1982). The peptide rises in response to nutritional repletion and has a high correlation with nitrogen metabolism and body weight (Phillips et al., 1978; Isley et al., 1983; Clemmons et al., 1985), thereby demonstrating its utility as an indicator of nutritional status. Long term studies of nutritional regulation of SMC/IGF 1 have not been reported. This study was undertaken to extend previous observations by investigating the impact of modest dietary restriction on SMC/IGF 1 during development, maturity and old age in the Lobund-Wistar (L-W) rat.

MATERIALS AND METHODS

Conventional male L-W rats sacrificed at ages corresponding to development (i.e. 2, 3, 5, 7 months), maturity (i.e. 18 months) and old age (i.e. 30 months) were used in this study. The animals were housed in plastic isolators which were opened to the local environment for introducing food and water. All rats were fed a steam-sterilized natural ingredient diet (L-485, Teklad, Madison, WI). The composition of the diet has been reported (Kellogg and Wostmann, 1969). Rats were fed either ad libitum or 12 grams of diet per day. This quantity represents approximately 70% of adult ad libitum intake and becomes restrictive for the male L-W rat at about 8 weeks of age. This level of restriction extends life span and does not interfere with normal reproduction (Snyder and Wostmann, 1987). See the chapter by Snyder and Wostmann in these proceedings for further details on the design of the study and the life span of L-W rats.

Blood was obtained between 9 A.M. and 9:30 A.M. after a 16 hour fast by heart puncture under light halothane anesthsia. The blood was allowed to clot for 30 minutes, placed on ice and centrifuged. Serum was collected and frozen at $-70^{\circ}C$ until analysis for serum SMC/IGF 1 concentration.

Concentration of SMC/IGF 1 was determined by raioimunoassay as previously described by Furlanetto et al. (1977) and modified by Copeland et al. (1980). Results are expressed as units per milliliter compared to a standard rat sera arbitrarily assigned a value of 1 unit/ml. The validity of the use of this heterologous assay in the rat has been previously demonstrated (Hurley et al., 1977).

A two way analysis of variance procedure was used to determine the effect of age, dietary restriction and their interaction on mean serum SMC/IGF 1. Because the results demonstrated a significant interaction of age and diet, Students T-test was used to determine the impact of dietary restriction on SMC/IGF 1 between groups at each age. Differences were considered significant at $p<0.05$, two-tailed test.

RESULTS

In developing animals, food intake was approximately 1.2 grams higher per 100 grams body weight in restricted compared to ad libitum animals at 2 months of age (10.0g vs 8.8g, respectively). This magnitude of difference between groups was also evident at 3 months and but not at 5 and 7 months. Food consumption was higher in restricted than ad libitum animals at 18 months and in old age.

Body weight of ad libitum animals was significantly higher at each age in comparison with restricted animals. Body weight for ad libitum vs restricted groups ranged from 153 $\pm$ 15.1 vs 119 $\pm$ 6.5 grams (mean $\pm$ SEM) at 2 months to 458 $\pm$ 5 vs 298 $\pm$ 3 grams at 18 months and 457 $\pm$ 13 vs 306 $\pm$ 7 grams at 30 months.

SMC/IGF 1 was significantly higher in ad libitum than in restricted animals at 5 months (1.53 $\pm$ .22 U/ml vs 1.00 $\pm$ .21 U/ml, mean $\pm$ SEM, $p<.001$, respectively) and in mature 18 month olds (1.41 $\pm$ .13 U/ml vs .94 $\pm$ .03 U/ml, $p<.05$, respectively). SMC/IGF 1 declined at 30 months in both groups, the difference in the peptide between dietary groups was not statistically significant.

DISCUSSION

Our data extend previous observations of nutritional regulation of SMC/IGF 1 (Grant et al., 1973; Phillips et al., 1978; Phillips et al., 1979; Prewitt et al., 1982; Isley et al., 1983; Clemmons et al., 1985) by characterizing the impact of modest dietary restriction over the life span in an experimental model. The impact of dietary restriction on SMC/IGF 1 varied by stage of the life cycle. During growth and development, when anabolism predominates, SMC/IGF 1 was significantly lower by 5 months in restricted as compared to ad libitum animals. The difference between groups was also evident among mature 18 month old animals, despite similar food consumption per 100 grams of body weight. The reduction in SMC/IGF 1 in growing animals consuming a restricted diet may represent an adaption where available nutrients are utilized for critical processes of survival by a mechanism which decreases SMC/IGF 1 synthesis. There is recent evidence that this decrease is due to a decrease in SMC/IGF 1 messenger RNA, but whether this is caused by decreased transcription of the gene or decreased mRNA stability is not known (Hurley et al., 1977; Elmer and Schalch, 1987). Thus the limited nutrient availability in restricted animals was reflected in lower SMC/IGF 1 up to maturity. Among aging 30 month old animals, however, SMC/IGF 1 was not different between dietary groups. This similarity in concentration of the peptide in aging animals may be partly a function of reduction in nutrient needs for anabolic processes.

The age-related decline in SMC/IGF 1 is well established. Florini and Roberts (1980) and Florini et al. (1981) showed that somatomedin-like growth factor declined with age from young (2 month) to middle aged (12-17 months) rats. Our data confirm this relationship between SMC/IGF 1 and age and suggest that during development and middle age, the normal rise in SMC/IGF 1 is blunted by modest dietary restriction. The reduction in serum SMC/IGF 1 with age appears to parallel the decline in nutrient requirements, specific mechanisms for this observation require further investigation.

Acknowledgements

Support provided by the Retirement Research Foundation, Park Ridge, IL and USPHS Grant HD08299.

REFERENCES

Clemmons DR, Underwood L, Dickerson R, Brown R, Hale L, Macpee R, Heizer W (1985). Use of plasma somatomedin-C/insulin like growth factor 1 measurements to monitor response to nutrition repletion in malnourished patients. Am J Clin Nutr 41:191-198.

Copeland KC, Underwood LE, Van Wyk JJ (1980). Induction of immunoreactive somatomedin-C in human serum by growth hormone. Dose response relationship and effect on chromatographic profiles. J Clin Endocrinol Metab 50:690-697.

Emler CA, Schalch D (1987). Nutritionally induced changes in hepatic insulin-like growth factor-1 (IGF-I) gene expression in rats. Endocrinology 120:832-834.

Florini J, Roberts S (1980). Effects of rat age in blood levels of somatomedin-like growth factors. J Gerontol 35:23-30.

Florini J, Haraed J,. Richman R, Weiss J (1981). Effects of rat age on serum levels of growth hormone somatomedins. Mech Ageing Dev 15:165-176.

Furlanetto R, Underwood L, Van Wyk JJ, D'Ercole AJ (1977). Estimation of somatomedin-C levels in normals and patients with pituitary disease by RIA. J Clin Invest 60:648-657.

Grant D, Hambley J, Becker D, Pimstone B (1973). Reduced sulfation factor in undernourished children. Arch Dis Child 48:596-600.

Hurley T, D'Ercole A, Handwerger S, Underwood L, Furlanetto R, Fellows R (1977). Ovine placental lactogen induces somatomedin: A possible role in fetal growth. Endocrinology 101:1635-1638.

Isley w, UnderwoodL, Clemmons D (1983). Dietary components that regulate serum SM-C concentration in humans. J Clin Invest 71:175-182.

Kellogg TS, Wostmann BS (1969). Stock diet for colony production of germfree rats and mice. Lab Anim Care 19:812-814.

Phillips L, Oranski A, Belosky D (1978). Somatomedin and Nutrition. IV: Regulation of somatomedin activity and growth cartilage activity by quantity and composition of diet in rats. Endocrinology 103:121-127.

Phillips L, Belosky D, Young H, Reichard L (1979). Somatomedin and Nutrition. VI: Somatomedin activity and somatomedin inhibitory activity in serum from normal and diabetic rats. Endocrinology 104:1519-1524.

Prewitt TE, D'Ercole AJ, Switzer BR, Van Wyk JJ (1982). Relationship of serum immunoreactive somatomedin-C to dietary protein and energy in growing rats. J Nutr 112:144-150.

Snyder DL, Wostmann BS (1987). Growth rate of the male germfree Wistar rat fed ad libitum or restricted natural ingredient diet. Lab Anim Sci 37:320-325.

CHANGES IN PANCREATIC HORMONES DURING AGING

Richard C. Adelman, PhD

Institute of Gerontology and
Department of Biological Chemistry
The University of Michigan
Ann Arbor, Michigan 48109

The ability to adapt to environmental challenge frequently is modified during aging. For example, there are several enzyme adaptations to stimulation by nutrients, drugs, hormones and so forth, which are known to change in time course and/or in magnitude of response as individuals of various species grow older (Obenrader et al, 1981). A specific case of such an enzyme adaptation that is altered during aging is that of hepatic glucokinase in response to treatment with glucose (Adelman, 1970). As male or female Sprague-Dawley rats age progressively from 2 to at least 24 months, the adaptive increase in enzyme activity following intragastric injection of glucose is progressively delayed in time of onset from three to approximately 10-12 hours, whereas magnitude of response is not altered.

This impaired adaptation of hepatic glucokinase to glucose during aging in Sprague-Dawley rats probably is not the consequence of functional decline in the hepatocyte. Each of at least the following experimental results contribute to such a conclusion. 1) Insulin is required for responsiveness by hepatic glucokinase to glucose (Sharma et al, 1963). Direct stimulation of hepatic glucokinase by intraperitoneal injection of insulin provokes an increase in enzyme activity for which neither time course nor magnitude of response is altered as rats age from 2 to at least 24 months (Adelman, 1970). However, this may not be regarded as conclusive evidence because the amounts of injected insulin required for adaptation by hepatic glucokinase are significantly greater than the concentration of insulin found in portal vein blood following enzyme adaptation in response to glucose. 2) Binding of insulin to its receptor on purified preparations of hepatic plasma membrane is not altered either in affinity or in number of binding sites as donor rats age beyond 12 months (Freeman et al, 1973). 3) Similar results were reported for other adaptive heapatic enzymes and their regulatory hormones, as reviewed previously (Sartin et al, 1980).

Changes in the adaptive regulation of hepatic glucokinase by glucose during aging in Sprague-Dawley rats probably reflect an imbalance of critical pancreatic hormones. The glucose-stimulated availability of insulin to liver is altered in a complex manner during aging. Glucose-stimulated secretion of insulin into portal vein blood in vivo occurs as a biphasic response in 2 month old rats (Gold et al, 1976). As rats age from 2 to 24 months, the initial phase of insulin response to glucose increases in magnitude, whereas the second phase is reduced in magnitude and requires greater time for onset of response.

It is tempting to associate the delayed second phase of insulin response to glucose to the delayed response by hepatic glucokinase during aging. However, any attempt to do so requires the assumption that the initial phase of insulin response to glucose is physiologically irrelevant to the adaptive increase in hepatic glucokinase activity. That such an assumption may be valid is indicated by the impaired ability of glucose to suppress secretion of glucagon during aging (Klug et al, 1979). Glucagon is the primary physiological antagonist to the action of insulin on liver. Furthermore, the molar ratio of immunoreactive glucagon to immunoreactive insulin is extremely high in portal vein blood of 24 month old rats during the initial hours of response to glucose. Therefore, it probably is the second phase of insulin response to glucose, the phase that is delayed in time of onset during aging, which accounts for the delayed adaptation of hepatic glucokinase in the aging Sprague-Dawley rat.

The complex insulin response to glucose detected in portal vein blood of aging Sprague-Dawley rats may reflect a change in distribution of heterogeneous populations of pancreatic islets of Langerhans (Kitahara, Adelman, 1979). Islet populations are readily separated on the basis of size. In 2 month old rats, the population of smallest islets predominates overwhelmingly, whereas the population of largest islets increases considerably by 24 months of age. Furthermore, the population of small islets loses its ability to respond rapidly to glucose in vitro, whereas insulin secretion by the large islets in vitro remains essentially intact, as donor rats age from 2 to 24 months. Therefore, the integrity of rapid insulin responsiveness to glucose during aging in the intact pancreas of Sprague-Dawley rats probably is maintained by the increasingly prominent role of the population of large islets. The mechanisms which underlie the change in distribution of islet populations in vivo and the impaired capability for insulin secretion in vitro by the small islets are unknown.

One possible mechanism by which the small islet population loses its ability to secrete insulin in vitro during aging of donor rats is the enhanced availability of pancreatic somatostatin, a potent endogenous inhibitor of glucose-stimulated insulin secretion. Two types of experimental results support such a possibility. 1) The rate of somatostatin secretion in vitro by small islets increases many fold as donor rats age from 2 to 24 months (Chaudhuri et al, 1983). 2) The impaired ability to stimulate insulin secretion in vitro by small islets from 24 month old rats is partially restored when these islets are incubated with antibodies to somatostatin (Chaudhuri et al, 1983).

Most recent efforts to understand the impact of aging on the regulation of somatostatin consider its biosynthesis, post-translational metabolism, and the biological potency of its metabolites with respect to control of insulin and glucagon secretion. One specific, pending example of such efforts entails an investigation of the distribution of immunoreactive somatostatin species in the pancreas of aging Lobund rats which are ad libitum fed and diet restricted.

REFERENCES

Adelman RC (1970). An age-dependent modification of enzyme regulation. J Biol Chem 245:1032-1035.

Chaudhuri M, Sartin JL, Adelman RC (1983). A role for somatostatin in the impaired insulin secretory response to glucose by islets from aging rats. J Geront 38:431-435.

Freeman C, Karoly K, Adelman RC (1973). Impairments in availability of insulin to liver in vivo and in binding of insulin to purified hepatic plasma membrane during aging. Biochem Biophys Res Commun 54:1573-1580.

Gold G, Karoly K, Freeman C, Adelman RC (1976). A possible role for insulin in the altered capability for hepatic enzyme adaptation during aging. Biochem Biophys Res Commun 73:1003-1010.

Kitahara A, Adelman RC (1979). Altered regulation of insulin secretion in isolated islets of different sizes in aging rats. Biochem Biophys Res Commun 87:1207-1213.

Klug TL, Karoly K, Adelman RC (1979). Altered regulation of pancreatic glucagon levels in rats during aging. Biochem Biophys Res Commun 89:907-912.

Obenrader MF, Sartin JL, Adelman RC (1981). Enzyme adaptation during aging. In Florini JR (ed): "Handbook of Biochemistry in Aging," Boca Raton: CRC Press, Inc., pp 263-267.

Sartin J, Chaudhuri M, Obenrader M, Adelman RC (1980). The role of hormones in changing adaptive mechanisms during aging. Fed Proc 39:3163-3167.

Sharma C, Manjeshwar R, Weinhouse S (1963). Effects of diet and insulin on glucose-adenosine triphosphate-phosphotransferases of rat liver. J Biol Chem 238:3840-3845.

NEUROENDOCRINOLOGY

Dietary Restriction and Aging, pages 169-180
© 1989 Alan R. Liss, Inc.

EVIDENCE THAT UNDERFEEDING ACTS VIA THE NEUROENDOCRINE SYSTEM TO INFLUENCE AGING PROCESSES

Joseph Meites

Department of Physiology, Michigan State University, East Lansing, Michigan 48824

INTRODUCTION

Underfeeding has been observed to inhibit development of pathology in many organs and tissues, to help maintain immune function, and to prolong lifespan in rats and mice (McCay et al., 1935; Everitt et al., 1982; Masoro, 1984; Weindruch et al., 1986). Tryptophan deficiency in the diet also was reported to prolong life in the rat (Segall and Timeras, 1976), but since these animals ate less food and lost body weight, the increase in lifespan may not be caused specifically by tryptophan deficiency. It is generally agreed that no specific dietary factor is responsible for the increased longevity of underfed rodents, but that this is due to reduced caloric intake (Masoro, 1984). Hypophysectomy also results in a marked reduction in food intake and, like food restriction, preserves many body organs and tissues (Everitt, 1983). However, hypophysectomized rats are "fragile" animals and, when maintained under normal laboratory conditions, do not live as long as intact ad libitum-fed rats (Meites et al., 1987). Only when placed in a protected environment (warm room, given food supplements, etc.) or receiving replacement doses of adrenal glucocorticoid hormones, may hypophysectomized rats live as long or longer than intact ad libitum-fed rats (Everitt, 1983).

We believe that the effects of reduced caloric intake in rodent species are exerted to a large extent by reducing neuroendocrine function, resulting in less "wear and tear," damage by hormones, toxins, and free radicals, all leading to preservation of the functional integrity of the neuro-endocrine system and the organs and tissues it regulates. It has been demonstrated that caloric restriction produces 1) a decrease in hypothalamic catecholamines (CA) necessary for release of hypothalamic hormones (Wurtman and Wurtman, 1983), 2) a decline in hypothalamic hypophysiotropic peptide hormones which act on the pituitary to release its hormones (Meites, 1970), and 3) reduced secretion of all pituitary and target gland hormones (Mulinos and Pomerantz, 1940; Meites, 1953; Campbell et al., 1977). Fig. 1 shows the effects of underfeeding and refeeding on five pituitary hormones in mature male rats. The condition that

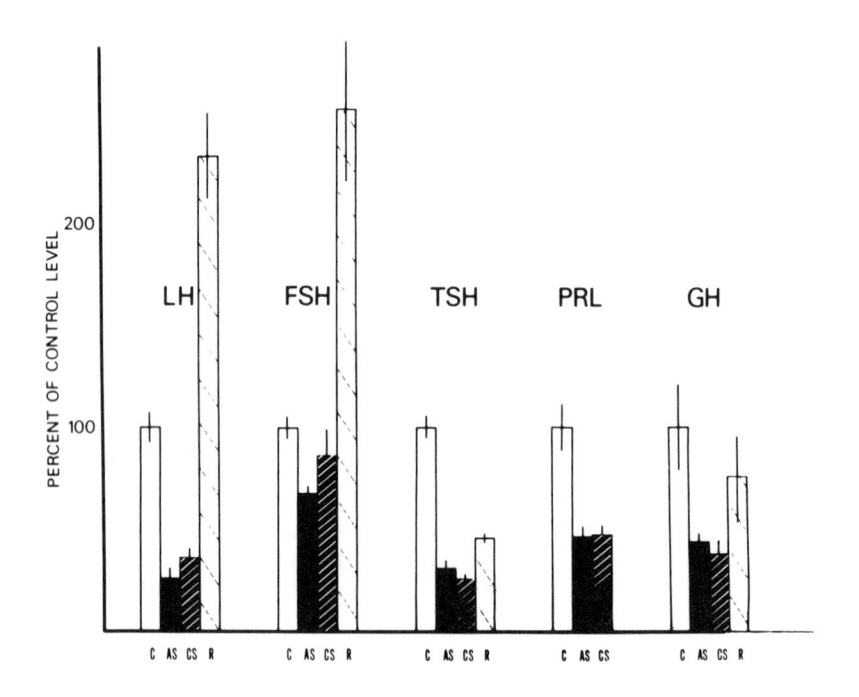

Figure 1. Effects of restricted food intake and refeeding on serum hormone levels. C = controls, fed ad libitum; AS = acutely starved, given no food for 7 days; CS = chronically starved, rats given 1/4 of ad libitum food intake for 2 weeks after 7 days sans food; R = refed, rats returned to ad libitum feeding after 14 days of chronic starvation. Note that rise in LH and FSH after refeeding was greater than levels in control rats (from Campbell et al., 1977).

ensues from severe underfeeding has sometimes been referred to as "pseudohypophysectomy" (Mulinos and Pomerantz, 1940). However, the effects of hypophysectomy on many body functions differ from those produced by dietary restriction, since hypophysectomy effectively removes all hormones from the pituitary and its target glands, resulting in lower body temperature, reduced blood glucose levels, decreased protein synthesis, much reduced capacity to respond to stressful stimuli, and other decrements of function that are much more severe than those usually present in animals on reduced food intake. The extent of the decrease in neuroendocrine function during underfeeding depends on the degree of dietary restriction--a greater reduction in food intake produces a larger decrease in neuroendocrine function.

EVIDENCE THAT DECREASED FOOD INTAKE ACTS VIA THE NEUROENDOCRINE SYSTEM TO PRODUCE SOME OF ITS EFFECTS

If reduced caloric intake alters body activities by reducing neuroendocrine function, then administration of hormones in doses sufficient to return hormone levels to those present in full-fed animals should counteract some of the effects of underfeeding. There is evidence that some effects of underfeeding can be overcome by hormone administration or by treatments that increase hormone secretion, as shown in the examples cited below.

Effects of Elevating Gonadotropins on Ovaries and Uterus of Underfed Rats

Several investigators have reported that the decrease in weight and function of the ovaries of underfed rats and guinea pigs can be prevented by injection of gonadotropic hormones (Marrian and Parkes, 1929; Stephens and Allen, 1941). We (Piacsek and Meites, 1967) observed that when mature female rats were provided with only 50% of the amount of food eaten by ad libitum-fed control rats for 31 days, they lost body weight, ceased to undergo estrous cycles within 14 to 21 days, and the ovaries and uterus became atrophic. When underfed animals were placed under constant light from the 21st to 31st days of underfeeding, they came into proestrus or estrus, ovarian weight increased, they exhibited well-developed follicles, and the uterus showed significant epithelial growth. Constant light is known to increase FSH secretion, stimulate follicle development and estrogen secretion, and increase uterine weight in rats (Negro-Vilar et al., 1968). Rats were also injected with epinephrine in corn oil from the 21st to 31st days of underfeeding, resulting in significant increases in ovarian and uterine weights and evidence of ovulation as indicated by the presence of corpora lutea in the ovaries. Intraventricular epinephrine and norepinephrine administration has been shown to stimulate gonadotropic hormone release in rabbits (Sawyer, 1952). It is logical to conclude, therefore, that the inhibitory effects of underfeeding on ovarian and uterine function by constant light or epinephrine were mediated by decreasing neuroendocrine operations.

Reversal of the Inhibitory Effects of Underfeeding on Growth of Carcinogen-Induced Mammary Cancers by Hormone Administration

One of the important effects of underfeeding is to reduce the onset of pathology, including development of tumors (Everitt, 1983; Masoro, 1984). This is believed to contribute importantly to prolongation of life in rodent species. In a study by Leung et al. (1983) (Fig. 2), female rats, 50-55 days of age, were given a single i.v. injection of an emulsion containing 5 mg of 7,12-dimethylbenz(a)-anthrace (DMBA). After 8-10 weeks, when each rat had at least one mammary tumor measuring more than 1 cm in diameter, they were placed in individual cages and given _ad libitum_ feeding (controls) or 50% of the amount of food eaten daily by the _ad libitum_-fed controls. Some of the half-fed rats were injected daily with estradiol benzoate (EB) to raise estrogen levels, haloperidol (HAL) to increase prolactin secretion, or both EB and HAL. Estrogen and prolactin are both essential for mammary tumor development in the rat (Meites, 1972). It can be seen that when half-fed rats were given HAL, it completely counteracted the effects of underfeeding on mammary tumor growth, whereas EB partially overcame the effects of underfeeding. When both EB and HAL were given to the underfed rats, mammary tumor growth was even greater than in the full-fed control rats.

The above results show that elevation of the two hormones known to be essential for mammary tumor

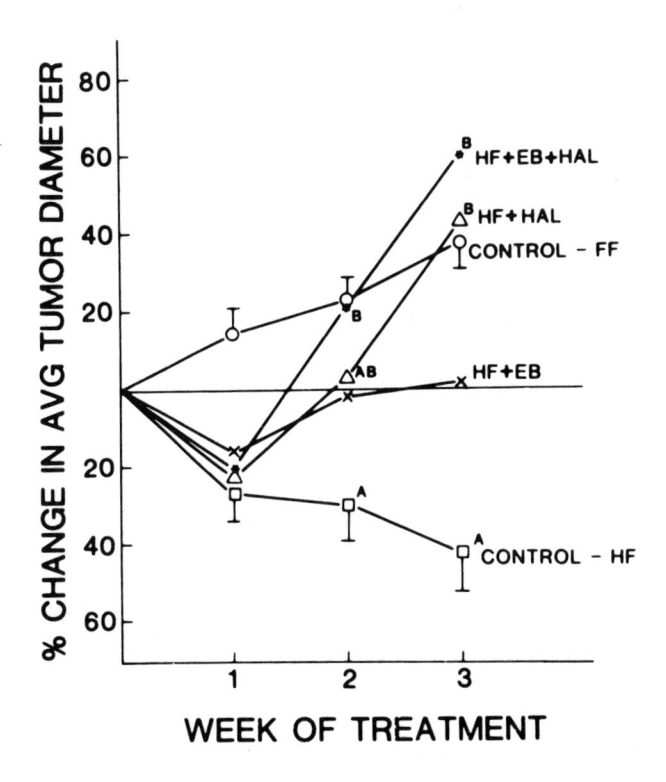

Fig. 2. Counteraction by injections of estradiol benzoate (DB) and/or haloperidol (HAL) of half-feeding (HF) on mammary tumor growth. FF = full-fed rats. Note that HAL alone completely overcame the effects of underfeeding, that EB partially counteracted the effects of underfeeding, and that the combination of EB and HAL produced even greater growth of mammary tumors in the half-fed than in the full-fed rats (from Leung et al., 1983).

development and growth in the rat completely counteracted regression of mammary tumor size induced by underfeeding. This demonstrates that the inhibitory effects of underfeeding on growth of carcinogen-induced mammary tumors in rats was achieved by reducing secretion of estrogen and prolactin. Serum prolactin levels were found to be significantly lower in the underfed than in the full-fed rats. Although estrogen levels were not measured, estrous cycles had ceased in the underfed rats, indicating that estrogen secretion was decreased. Whether inhibition of development and growth of other tumors by dietary restriction can be counteracted by elevating hormone levels remains to be determined. Also, it remains to be seen whether the effects of more prolonged or chronic underfeeding on mammary tumors can be overcome by continuous elevation of estrogen and prolactin.

Evidence that Short-Term Underfeeding Followed by Ad Libitum Feeding Can Temporarily Improve Reproductive Function

In a recent experiment, we (Quigley et al., 1987) placed female rats 5-6 months or 15-16 months old on 50% normal food intake for a period of 10 weeks, followed by ad libitum feeding for 16 weeks. Initially, 100% of the young rats were exhibiting estrous cycles and only about 40% of the older rats were cycling. During the 10 weeks of underfeeding, all young and older rats ceased to cycle (Fig. 3). After placement on ad libitum feeding, all rats rapidly regained body weight, and 100% of both young and old rats resumed cycling. All the previously underfed young rats continued to cycle for about 10 weeks followed by a gradual decline in the number of cycling rats, but a larger number continued to cycle than in the continuously ad libitum-fed young controls. All the older rats initially cycled and then showed a more rapid decline in cycling after being placed on ad libitum feeding than the young rats, but continued cycling in greater numbers than in the continuously ad libitum-fed old controls. In a subsequent experiment, we (Quigley and Meites, unpublished) found that when young and middle-aged female rats were fed 50% of normal food intake for 10 weeks, followed by ad libitum feeding, the rise in LH upon challenge by estrogen and progesterone after ovariectomy (positive feedback) was greater than in the continuously ad libitum-fed controls. Note that in Fig. 1, underfeeding followed by refeeding also resulted in a large rebound of both LH and FSH, to levels greater than initially present in these rats. The increase in ability to release LH upon refeeding after a period of underfeeding may be due in part to the reduction in estrogen secretion that occurs during underfeeding, resulting in less damage to hypothalamic neurons involved in regulating estrous cycles. Chronic estrogen action has

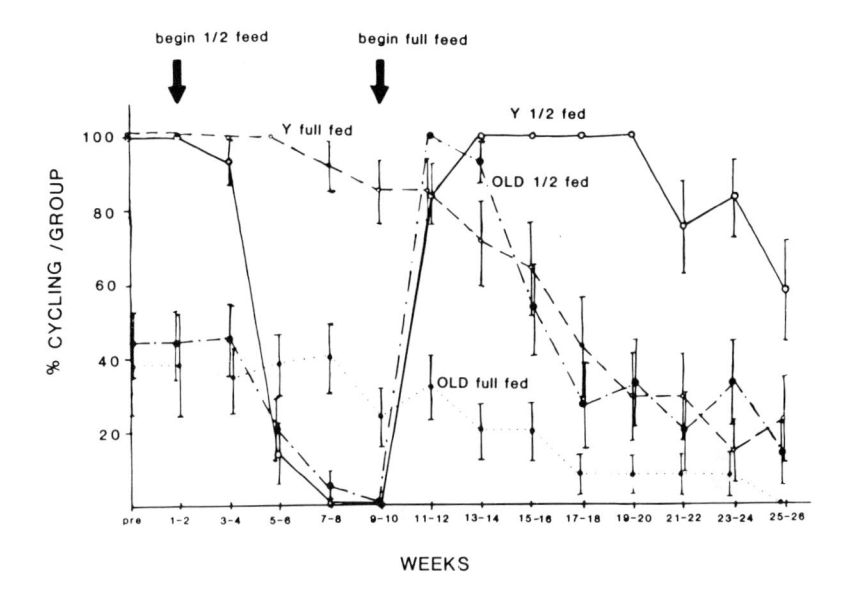

Fig. 3. Effects of half-feeding for 10 weeks followed by full-feeding for 16 weeks on per cent of young (Y) and old rats undergoing estrous cycles. Note that 100% of both young and old rats ceased cycling during underfeeding, and that 100% of both young and old rats temporarily resumed cycling upon refeeding. Thereafter cycling gradually declined in both young and old previously half-fed rats, but the per cent of these rats cycling always remained above the full-fed control rats. Rats normally show a decline in cycling beginning at about 7 months of age (from Quigley et al., 1987).

been shown to damage hypothalamic neurons (Brawer et al., 1978). The decreased activity of the neuroendocrine system during underfeeding and possibly a reduction in free radical damage may also have contributed to the temporary improvement in reproductive function upon refeeding.

DISCUSSION

The studies cited here indicate that underfeeding inhibits ovarian function and mammary tumor growth by depressing neuroendocrine function, resulting in a decrease in secretion of the hormones necessary to maintain ovarian function and mammary tumor growth. When the essential hormones were elevated during underfeeding, ovarian function and mammary tumor growth resumed. Whether other effects of underfeeding can be counteracted by elevating neuroendocrine activity remains to be determined. It is also important to study the extent to which the effects of chronic underfeeding can be overcome by elevating hormone levels.

In the experiment on the effects of 10 weeks of underfeeding followed by refeeding on estrous cycles, it is not entirely clear why the 60% of the older rats that had ceased to cycle prior to underfeeding resumed cycling upon refeeding. It appears that underfeeding not only permitted estrous cycles to continue in the young rats, but in addition resulted in a rejuvenation of the mechanisms regulating estrous cycles in the older rats. Although there may have been less damage by estrogen and free radicals during underfeeding, these alone do not explain the reinitiation of cycles in the older rats.

Little is yet known of immune-neuroendocrine interactions during underfeeding. Underfeeding has been reported to lessen the age-related decline in thymus dependent immunological function in mice (Cheney et al., 1983). Even when underfeeding was begun in mid-adulthood, it helped to maintain immune competence in mice (Weindruch et al., 1982). This may partially account for the reduced incidence of disease in rats and mice on restricted caloric intake, and may also contribute importantly to prolongation of lifespan. Old rats and mice, as well as humans, usually die because of disease (Brody and Brock, 1985).

What is the role of hormones in reducing the decline in immune function during underfeeding? In general, GH, PRL, and thyroid hormones have been shown to promote growth and function of the thymus, spleen, and lymph glands, whereas adrenal glucocorticoid and gonadal hormone produce opposite effects (Comsa et al., 1982). Although all of these hormones are lowered by mild or moderate degrees of restricted caloric intake, the reductions in adrenal glucocorticoid and gonadal hormones may be particularly significant for immune function. The reduced inhibitory effects by these hormones may permit the immune system to operate at a higher level. The anti-immune actions of adrenal glucocorticoid hormones are well established. As for gonadal hormones, it was reported that when old male rats were castrated, thymus size and function returned to youthful levels, whereas testosterone administration depressed thymus size and function (Greenstein et al., 1986). The decrease in thymus size at puberty is well known to be due to the increase in gonadal hormone secretion at this time. Thus part of the mechanism by which underfeeding helps to maintain immune function may involve a lowering of secretion of adrenal glucocorticoid and gonadal hormones. Restricted dietary intake may also operate via non-hormonal mechanisms to promote immune competency, but these remain to be clarified.

In ad libitum-fed rodents immune competency declines with age. There is now convincing evidence that this decline is at least partially due to lower secretion of GH and thyroid hormones. Both GH secretion (Sonntag et al., 1980; Takahashi, 1987) and thyroxine (T4) (Huang et al., 1980) have been shown to decline in aging rats. Administration of GH was recently demonstrated to increase size and function of the thymus in aging dogs (Monroe et al., 1987) and rats (Kelley et al., 1986). Fabris (1973) observed that thyroidectomy depressed the humoral response to antigens in young adult rats, whereas thyroxine completely restored the lymphoid system of thyroid-deprived animals. It can be concluded, therefore, that the neuroendocrine system is intimately involved in the normal decline in immune competency during aging, as well as in the underfeeding effects on the immune system.

REFERENCES

Brawer GR, Naftolin J, Martin J, Sonnenschein C (1978). Effects of a single injection of estradiol valerate on the hypothalamic arcuate nucleus and on reproductive function in the female rat. Endocrinology 103:501-512.

Brody JA, Brock DB (1985). Epidemiologic and statistical characterisics of the United States elderly population. In Finch CE, Schneider EL (eds), "Handbook of the Biology of Aging 2nd Edn," New York: Van Nostrand Reinhold Co, pp 3-26.

Campbell, GA, Kurcz M, Marshall S, Meites J (1977). Effects of starvation on serum levels of follicle stimulating hormone, luteinizing hormone, thyrotropin, growth hormone and prolactin; response to LH-releasing hormone and thyrotropin releasing hormone. Endocrinology 100:580-587.

Cheney KE, Liu RK, Smith GS, Meredith PJ, Mickey MR, Walford RL (1983). The effect of dietary restriction of varying duration on survival, tumor patterns, immune function, and body temperature in B10C3F1 female mice. J Geront 18:427-435.

Comsa J, Leonhardt H, Wekerle H (1982). Hormonal coordination of the immune response. Rev Physiol Biochem Pharmacol 92:115-191.

Everitt AV (1976). Hypophysectomy and aging in the rat. In Everitt AV, Burgess JA (eds): "Hypothalamus, Pituitary and Aging," Springfield, IL: Chas C Thomas, pp 68-85.

Everitt AV (1983). Pacemaker mechanisms in aging and the diseases of aging. In Blumenthal HT (ed): "Handbook of Diseases of Aging," New York: Van Nostrand Reinhold Co, pp 93-132.

Everitt AV, Porter BD, Wyndham JR (1982). Effects of caloric intake and dietary composition on the development of proteinuria, age-associated renal disease, and longevity in the male rat. Gerontology 28:168-175.

Fabris N (1973). Immunodepression in thyroid-deprived animals. Clin Exp Immunol 15:601-611.

Greenstein BD, Fitzpatrick FTA, Adcock IM, Kendall MD, Wheeler MJ (1986). Reappearance of the thymus in old rats after orchidectomy: inhibition of regeneration by testosterone. J Endocrinology 110:417-422.

Huang HH, Steger RW, Meites J (1980). Capacity of old versus young male rats to release thyrotropin (TSH), thyroxine (T4) and triiodothyronine (T3) in response to different stimuli. Exper Aging Res 6:3-11.

Kelley KW, Brief S, Westly HJ, Novakofski J, Bechtel PJ, Simon J, Walker EB (1986). GH3 pituitary adenoma cells can reverse thymic aging in rats. Proc Natl Acad Sci 83:5663-5667.

Leung FC, Aylsworth CF, Meites J (1983). Counteraction of underfeeding-induced inhibition of mammary tumor growth in rats by prolactin and estrogen administration. Proc Soc Exp Biol Med 173:159-163.

Marrian GF, Parkes AS (1929). The effects of anterior pituitary preparations administered during dietary anestrus. Proc Roy Soc (London) B105:248-258.

Masoro EJ (1984). Nutrition as a modulator of the aging process. The Physiologist 27:98-101.

McCay CM, Crowell LA, Maynard J (1935). The effect of retarded growth upon the length of lifespan and upon the ultimate body size. J Nutr 10:63-79.

Meites J (1953). Relation of nutrition to endocrine-
reproductive functions. Iowa State College Journal of
Science 28:19-44.
Meites J (1970). Modification of synthesis and release of
hypothalamic releasing factors induced by exogenous
stimuli. In Martini L, Meites J (eds): "Neurochemical
Aspects of Hypothalamic Function," New York: Academic
Press, pp 1-44.
Meites J, Goya R, Takahashi S (1987). Why the neuro-
endocrine system is important in aging processes. Exper
Geront 22:1-15.
Monroe WE, Roth JA, Grier RL, Arp LH, Naylor PH (1987).
Effects of growth hormone on the adult canine thymus.
Thymus 9:173-187.
Mulinos MG, Pomerantz L (1940): Pseudohypophysectomy, a
condition resembling hypophysectomy, produced by
malnutrition. J Nutr 19:493-504.
Piacsek BE, Meites J (1967). Reinitiation of gonadotropin
release in underfed rats by constant light or
epinephrine. Endocrinology 81:535-541.
Quigley K, Goya R, Meites J (1987). Rejuvenating effects of
10-week underfeeding period on estrous cycles in young
and old rats. Neurobiology of Aging 8:225-232.
Sawyer CH (1952). Stimulation of ovulation in the rabbit by
the intraventricular injection of epinephrine or
norepinephrine. Anat Rec 112:385.
Segall PE, Timeras PS (1976). Patho-physiologic findings
after chronic tryptophan deficiency in rats: a model for
delayed growth and aging. Mech Aging Dev 5:109-124.
Sonntag WE, Steger RW, Forman LJ, Meites J (1980).
Decreased pulsatile release of growth hormone in old male
rats. Endocrinology 107:1875-1879.
Stephens DJ, Allen WM (1941). The effect of refeeding and
of the administration of a pituitary extract on the
ovaries of undernourished guinea pigs. Endocrinology
28:580-584.
Takahashi S, Gottschall PE, Quigley KL, Goya RG, Meites J
(1987). Growth hormone secretory patterns in young,
middle-aged, and old female rats. Neuroendocrinology
46:137-142.
Weindruch R, Gottesman SRS, Walford RL (1982). Modification
of age-related immune decline in mice dietarily restricted from or after adulthood. Proc Natl Acad Sci USA
79:898-902.

Weindruch R, Walford R, Fliegiel S, Guthrie D (1986). The retardation of aging in mice by dietary restriction: longevity, cancer, immunity and lifetime energy intake. J Nutr 116:641-654.

Wurtman RJ, Wurtman JJ (1983). "Nutrition and the Brain, Vol. 2." New York: Raven Press, pp 177-181.

ADENOHYPOPHYSIAL CHANGES IN CONVENTIONAL, GERM FREE AND FOOD-RESTRICTED AGING LOBUND-WISTAR RATS. A HISTOLOGIC, IMMUNOCYTOCHEMICAL AND ELECTRON MICROSCOPIC STUDY

Kalman Kovacs, Nancy Ryan, Toshiaki Sano, Lucia Stefaneanu, Gezina Ilse and Sylvia L. Asa

Department of Pathology, St. Michael's Hospital, University of Toronto, Toronto, Ontario, Canada

A germ free state and dietary restriction cause substantial alterations in the morphology and functional activity of several organs, including the endocrine glands. Thus, the detailed examination of tissues of germ free and food-restricted animals, using various morphologic and biochemical methods, is fully justified. Credit should be given to the Lobund Laboratory for initiating and organizing a multidisciplinary project involving a large number of investigators from several centers. Conventional, germ free and food-restricted Lobund-Wistar rats were kept at the Lobund Laboratory, University of Notre Dame, until the animals became old. After they died or were killed, the various organs were made available for study. We were asked to investigate the morphologic features of the pituitary glands.

In any study focusing on pituitary structure and function of old rats, changes which occur spontaneously in the course of aging have to be considered. Spontaneous pituitary adenomas are known to develop commonly in various strains of old rats; the frequency varies from 1 to 96% (Wolfe et al 1938; Saxton and Graham 1944; Wolfe and Wright 1947; Ito et al 1972; McComb et al 1984; McComb et al 1985a). The morphologic alterations in pituitaries of old rats have been extensively studied recently using histologic, immunocytochemical and electron microscopic techniques (Kovacs et al 1977; McComb et al 1981; Lee et al 1982; Trouillas et al 1982; Kovacs et al 1983; McComb et al 1984). The most often occurring tumors are adenomas composed of lactotrophs. These tumors contain immunoreactive prolactin (PRL), possess a characteristic ultrastructure and are accompanied by various degrees of hyperprolactinemia. Tumors other than PRL-producing adenomas have also been described (Berkvens et al 1980). These adenomas can be plurihormonal producing growth hormone (GH), PRL, adrenocorticotroph hormone (ACTH) and/or thyrotrop hormone (TSH). Some tumors show no specific immunocytochemical and electron microscopic features and are difficult to classify (Kovacs et al 1983; Kovacs and Horvath 1986). In a recent histologic, immunocytochemical and fine structural study, McComb et al (1985b) reported the development of follicle stimulating hormone (FSH) and luteinizing hormone (LH) producing gonadotroph adenomas in pituitaries of aging Sprague-Dawley rats. The existence of these tumors was suspected but not proven before the introduction of immunocytochemical and electron microscopic methods (Clifton 1959; Griesbach and Purves 1960; Griesbach 1967). Since considerable differences exist in incidence, morphologic characteristics and ultrastructure of pituitary adenomas found in different strains of rats, it is imperative that every meaningful study dealing with pituitary adenomas must include a large number of control animals. We report here the histologic, immunocytochemical and electron microscopic findings in adenohypophyses of conventional, germ free and food-restricted old Lobund-Wistar rats.

MATERIALS AND METHODS

Male Lobund-Wistar rats, from 6 months to 45 months of age, were used. The animals, kept at the Lobund Laboratory, University of Notre Dame, were killed by exsanguination in halothane anesthesia. Pituitaries, embedded for morphologic studies, were shipped to Toronto. Maintenance and treatment details, autopsy findings, changes in other organs, blood hormone levels, etc. are described in other chapters of this book.

For light microscopy, pituitaries were fixed in 10% formalin and embedded in paraffin. Sections of 4-6 μm thickness were stained with hematoxylin-eosin, the PAS technique and for the demonstration of the reticulin fiber network, with the Gordon-Sweet method.

For immunocytochemistry, paraffin sections of 4-6 μm thickness were used. For the demonstration of various adenohypophysial hormones, the avidin-biotin-peroxidase complex technique was applied as reported elsewhere (Hsu et al 1981a and b; McComb et al 1984). The source of various antisera, their dilution, exposure times, control procedures have been described in other papers (McComb et al 1984; McComb et al 1985a).

For electron microscopy, glutaraldehyde fixed, osmicated

pituitary tissues were embedded in Araldite. Semithin sections were stained with toluidine blue and appropriate areas selected for fine structural study. Ultrathin sections were cut on an LKB-Huxley ultramicrotome, stained with uranyl acetate and lead citrate and investigated with a Philips 410-LS electron microscope.

RESULTS

The morphologic study focused primarily on adenomas. In addition, non-tumorous adenohypophyses were investigated. The posterior pituitaries examined only in a few cases by histology, showed no major abnormalities.

Adenomas. The adenoma types and their frequency are shown in the following Table:

Frequency of pituitary adenomas

	Number of rats	With lactotroph adenoma number and percentage	With gonadotroph adenoma number and percentage	With other pituitary adenoma number and percentage
Conventional Food unrestricted	30	0 0	11 36.6	1* 3.3
Germ free Food unrestricted	26	2 7.6	3 11.5	0 0
Conventional Food restricted	41	3 7.3	7 17.0	0 0
Germ free Food restricted	35	1 2.8	6 17.1	1** 2.8

* thyrotroph adenoma
** PRL and TSH-producing adenoma

PRL-producing adenomas were uncommon. Gonadotroph adenomas, however, occurred frequently in the conventional control group which received food ad libidum. In contrast to this group, the incidence of gonadotroph adenomas was less in germ free and food-restricted animals. One adenoma from the control group contained only TSH. Another adenoma from a germ free, food-restricted rat was immunoreactive for PRL and TSH. The question of whether these two hormones were present in the cytoplasm of the same cells or two separate cell types was not investigated. Adenoma types, containing adenohypophysial hormones other than PRL, FSH, LH or TSH were not encountered in the present study.

Lactotroph and gonadotroph adenomas exhibited characteristic morphologic features described in detail in previous publications (Kovacs et al 1977; McComb et al 1981; Kovacs et al 1983; McComb et al 1984; McComb et al 1985a and b).

Lactotroph adenomas consisted of chromophobic, slightly acidophilic, PAS negative cells which contained immunoreactive PRL shown by the avidin-biotin-peroxidase complex technique. The tumors exhibited a diffuse pattern and groups of tumor cells were interspersed with dilated capillaries filled with conglomerated erythrocytes. The reticulin network was disrupted in the tumors. By electron microscopy as seen in previous studies (Kovacs et al 1977; McComb et al 1981; Kovacs et al 1983; McComb et al 1984, 1985a and b), lactotrophs possessed prominent rough endoplasmic reticulum membranes forming, in some places, concentric whorls, called Nebenkerns. They had conspicuous Golgi complexes and a few spherical or irregular, evenly electron dense membrane-bound secretory granules measuring 100-300 nm. Extrusion of secretory granules at the basal and lateral cell surfaces was a common finding.

Gonadotroph adenomas exhibited a diffuse or trabecular pattern and were composed of basophilic or chromophobic, strongly PAS positive cells. In contrast to lactotroph adenomas which occupied a large area of the anterior lobe and were not sharply demarcated in several places, gonadotroph adenomas were usually small, had a well-defined border and were surrounded by condensed reticulin fibers. The adenoma cells showed varying immunopositivity for FSH and strong immunopositivity for LH. The small nodules contained irregularly arranged reticulin fibers and were often interspersed with a few non-tumorous adenohypophysial cells containing immunoreactive GH, PRL, ACTH or TSH. In some cases the adenomas were so small that it was uncertain whether they were bona fide tumors or represented hyperplastic nodules. By electron microscopy, some adenoma cells showed a resemblance to non-tumorous gonadotrophs, whereas others exhibited different ultrastructure and based on electron microscopic appearance, their origin could not be recognized. Some adenoma cells were densely granulated, whereas others contained only a few secretory granules. The rough endoplasmic reticulum membranes and Golgi complexes were conspicuous in the well-granulated tumor cells, whereas they were scanty in adenomas consisting of small, sparsely granulated, less differentiated cells. The secretory granules of well-granulated cells were spherical, oval or irregular, varied in electron density and measured 50-250 nm. No exocytosis of secretory granules was evident. Some adenoma cells contained many large lysosomal bodies. One tumor in the sella region, harvested from a rat of the conventional food unrestricted group, exhibited the characteristic histologic features of meningioma.

Non-tumorous adenohypophyses. In many adenohypophyses of aging rats of various strains, lactotrophs were noted to be numerous (Kovacs et al 1980; McComb et al 1986). They formed groups of various sizes, were easily recognizable on immunostained sections and represented a prominent cell type of the anterior pituitary. In the present study, no lactotroph hyperplasia was found. In some adenohypophyses, gonadotrophs were conspicuous. They were large and were seen singly or forming small groups of ovoid cells, showing strong cytoplasmic PAS positivity and a prominent negative Golgi image. Immunocytochemistry revealed FSH and LH in their cytoplasm. Some gonadotrophs contained a large, irregular nucleus exhibiting pleomorphic features. GH, ACTH and TSH-containing cells showed no major abnormalities. Electron microscopy revealed several lactotrophs exhibiting ultrastructural features indicative of active secretion. They possessed prominent rough endoplasmic reticulum membranes as well as Golgi apparatus, and a variable number of spherical or irregular secretory granules measuring 100-350 nm. The gonadotrophs contained well developed rough endoplasmic reticulum membranes exhibiting varying degrees of dilatation, prominent Golgi apparatus and, in some cells, several large lysosomes. They were usually densely granulated often containing two populations of secretory granules, measuring 150-200 nm and 300-400 nm, respectively. So-called light bodies were also encountered.

There was no appreciable difference in cell morphology among the various animal groups.

DISCUSSION

The present study clearly indicates that major morphologic changes can often be found in the adenohypophyses of aging Lobund-Wistar rats. Two morphologically distinct tumor types, lactotroph and gonadotroph adenomas, were identified. In addition, one TSH-producing adenoma and one PRL and TSH-producing adenoma were found.

Lactotroph adenomas showed the characteristic morphology as described in previous papers (Kovacs et al 1977, 1983; McComb et al 1981, 1984, 1985a) and their appearance did not differ considerably among various animal groups. Compared to several other rat strains, they seemed to be uncommon in aging Lobund-Wistar rats. Gonadotroph adenomas producing FSH and LH were disclosed with greater frequency; they exhibited typical morphologic features, reported elsewhere (McComb et al 1985b). It was remarkable that, compared to conventional control groups which received food ad libitum, adenoma incidence was markedly reduced in germ free as well as food-restricted groups and in those rats in which the germ free state was accompanied by a restricted food intake.

The pathogenesis of pituitary adenomas is obscure. A primary hypothalamic abnormality has been proposed to be the cause of tumor formation in rats harboring PRL-producing lactotroph adenoma (Sarkar et al 1982; Sarkar et al 1983a,b). It was claimed that decrease in hypothalamic concentration of dopamine, the main PRL release inhibiting factor, or defect of dopamine release or inhibition of its transport to the adenohypophysis relieves lactotrophs of suppression, resulting in increased PRL secretion, lactotroph hyperplasia and adenoma formation (Meites 1982; Sarkar et al 1982; Sarkar et al 1983 a,b). It was also suggested that pituitary neovascularization plays a role in the development of lactotroph adenomas. According to Elias and Werner (1984) and Schechter et al (1987), non-portal vessels may grow into the adenohypophysis and contribute to adenohypophysial blood supply. These newly developed vessels carry non-hypothalamic, thus dopamine-poor blood leading to activation of lactotrophs, excessive PRL release, hyperprolactinemia, lactotroph proliferation and subsequently adenoma formation (Elias and Weiner 1984; Elias and Weiner 1987; Schechter et al 1987). Our previous finding (Kovacs et al 1980) of lactotroph hyperplasia, noted in the adenohypophyses of old rats of various strains, is consistent with both of these hypotheses. However, no lactotroph hyperplasia was observed in the adenohypophyses of aging male Lobund-Wistar rats. It should also be mentioned that the present investigation was not designed to shed light on the pathogenesis of lactotroph adenomas.

No plausible hypotheses were forwarded previously to explain the development of FSH and LH-producing gonadotroph adenomas in the pituitaries of aging rats. One possibility could be that old rats are hypogonad and because of lack of negative feedback effect of target gland hormones, pituitary gonadotrophs are stimulated, undergo replication, hyperplasia and neoplastic transformation. Decreased gonadal function is not supported by adenohypophysial morphology, since gonadectomy cells, markers of increased secretory activity of gonadotrophs, were not seen in the pituitaries. It remains to be seen whether a primary hypothalamic defect resulting in sustained gonadotroph hormone-releasing hormone (GnRH) oversecretion might have contributed to stimulation of pituitary gonadotrophs, their hyperplasia and subsequent formation of adenomas. Hypothalamic releasing hormones, such as growth hormone-releasing hormone (GRH) and corticotrop hormone-releasing hormone (CRH), are known to cause proliferation of the respective pituitary cell types (Billestrup et al 1986; Gertz et al 1987). Obviously, several other mechanisms, such as abnormal receptors, stimulation by various growth factors, oncogenes, have to be considered and it is clear that more work is required to provide an appropriate interpretation for the development of gonadotroph adenomas. The reduced incidence of gonadotroph adenomas in germ free and food-restricted rats, as well as in those which were in a grem free state and consumed subnormal amounts of food, was a striking finding in the present study. The cause of this decreased frequency has yet to be elucidated.

SUMMARY

The pituitaries of aging male Lobund-Wistar rats were investigated by histology, immunocytochemistry and electron microscopy. In contrast to previous studies undertaken on aging rats of various strains, no lactotroph hyperplasia was found in the adenohypophysis of the present study. Lactotroph adenomas producing PRL were infrequent, whereas gonadotroph adenomas containing FSH and LH were common. The frequency of gonadotroph adenomas was decreased in germ free as well as food-restricted rats and in those in which the germ free state was associated with reduced food intake.

ACKNOWLEDGEMENT

This work was supported in part by Grant MT-6349 awarded by the Medical Research Council. The authors wish to thank Dr. Eva Horvath for her help in evaluating the electron microscopic results and Mrs. Wanda Wlodarski for excellent secretarial work.

REFERENCES

Billestrup N, Swanson LW, Vale W (1986) Growth hormone-releasing factor stimulates proliferation of somatotrophs in vitro. Proc Natl Acad Sci USA 83:6854.

Berkvens JM, Van Nesselrooy JH, Kroes R (1980) Spontaneous tumors in the pituitary gland of old Wistar rats. A morphologic and immunocytochemical study. J Pathol 130:179.

Clifton KH (1959) Problems in experimental tumorigenesis of the pituitary gland, gonads, adrenal cortices and mammary glands: a review. Cancer Res 19:2.

Elias KA, Weiner RI (1984) Direct arterial vascularization of estrogen-induced prolactin secreting anterior pituitary tumors. Proc Natl Acad Sci USA 81:4549.

Elias KA, Weiner RI (1987) Inhibition of estrogen-induced anterior pituitary enlargement and arteriogenesis by bromocriptine in Fischer 344 rats. Endocrinology 120:617.

Gertz BJ, Conteras LN, McComb DJ, Kovacs K, Tyrrell JB, Dallman MF (1987) Chronic administration of corticotropin-releasing factor increases pituitary corticotroph number. Endocrinology 120:381.

Griesbach WE (1967) Basophil adenomata in the pituitary glands of a 2-year-old male Long-Evans rat. Cancer Res 27:1813.

Griesbach WE, Purves HD (1960) Basophil adenomata in the rat hypophysis after gonadectomy. Br J Cancer 14:49.

Hsu SM, Raine L, Fanger H (1981a) A comparative study of the peroxidase-antiperoxidase method and an avidin-biotin complex method for studying polypeptide hormones with radioimmunoassay antibodies. Am J Clin Pathol 75:734.

Hsu SM, Raine L, Fanger H (1981b) The use of antiavidin antibody and avidin-biotin-peroxidase complex in immunoperoxidase technics. Am J Clin Pathol 75:816.

Ito A, Moy P, Kaunitz H, Kortwright K, Clarke S, Furth J, Meites J (1972) Incidence and character of spontaneous pituitary tumors in strain CR and W/Fu male rats. J Natl Cancer Inst 49:701.

Kovacs K, Horvath E (1986) Tumors of the pituitary gland. Atlas of tumor pathology, Second Series, Fascicle 21,

Armed Forces Inst. of Pathology, Washington, D.C.
Kovacs K, Horvath E, Ilse RG, Ezrin C, Ilse D (1977) Spontaneous pituitary adenomas in aging rats. A light mircoscopic, immunocytochemical and fine structural study. Beitr Pathol 161:1.
Kovacs K, Ilse G, Ryan N, McComb DJ, Horvath E, Chen HJ, Walfish PG (1980) Pituitary prolactin cell hyperplasia. Hormone Res 12:87.
Kovacs K, McComb DJ, Horvath E (1983) Subcellular investigation of experimental and human pituitary adenomas. Neuroendocrinol Perspect 2:251.
Lee AK, DeLellis RA, Blount M, Nunnemacher G, Wolfe HJ (1982) Lab Invest 47:595.
McComb DJ, Hellmann P, Kovacs K, Scott D, Evans WS, Burdman JA, Thorner MO (1985a) Spontaneous sparsely granulated prolactin-producing pituitary adenomas in aging rats: a prospective study of the effect of bromocriptine. Neuroendocrinology 41:201.
McComb DJ, Hellmann P, Thorner MO, Scott D, Evans WS, Kovacs K (1986) Morphologic effects of bromocriptine on spontaneously occurring pituitary prolactin cell hyperplasia in old Long-Evans rats. Am J Pathol 122:7.
McComb DJ, Kovacs K, Beri J, Zak F (1984) Pituitary adenomas in old Sprague-Dawley rats: A histologic, ultrastructural, and immunocytochemical study. J Natl Cancer Inst 73:1143.
McComb DJ, Kovacs K, Beri J, Zak F, Milligan JV, Shin SH (1985b) Pituitary gonadotroph adenomas in old Sprague-Dawley rats. J Submicrosc Cytol 17:517.
McComb DJ, Ryan N, Horvath E, Kovacs K, Nagy E, Berczi I, Domokos I, Laszlo FA (1981) Five different adenomas derived from the rat adenohypophysis: immunocytochemical and ultrastructural study. J Natl Cancer Inst 66:1103.
Meites J (1982) Changes in neuroendocrine control of anterior pituitary function during aging. Neuroendocrinology 34:151.
Sarkar DK, Gottschall PE, Meites J (1982) Damage to hypothalamic dopaminergic neurons is associated with development of prolactin-secreting pituitary tumors. Science 218:684.
Sarkar DK, Gottschall PE, Meites J (1983a) Relation of the neuroendocrine system to development of prolactin-secreting pituitary tumors. In: Neuroendocrinology of aging. Ed. by Meites J, Plenum, New York, pp. 353.
Sarkar DK, Miki N, Meites J (1983b) Failure of prolactin short loop feedback mechanism to operate in old as compared to young female rats. Endocrinology 113:1452.
Saxton JA, Graham JB (1944) Chromophobe adenoma-like

lesions of the rat hypophysis. Frequency of the spontaneous lesions and characteristics of growth of homologous intraocular transplants. Cancer Res 4:168.

Schechter J, Ahmad N, Elias K, Weiner R (1987) Estrogen-induced tumors: changes in the vasculature in two strains of rat. Am J Anat 179:315.

Trouillas J, Girod C, Claustrat B, Cure M, Dubois MP (1982) Spontaneous pituitary tumors in the Wistar/Furth/Ico rat strain. Am J Pathol 109:57.

Wolfe JM, Bryan WR, Wright AW (1938) Histologic observations on the anterior pituitaries of old rats with particular reference to the spontaneous appearance of pituitary adenomata. Am J Cancer 34:352.

Wolfe JM, Wright AW (1947) Cytology of spontaneous adenomas in the pituitary gland of the rat. Cancer Res 7:759.

THE MORPHOLOGY OF ADENOHYPOPHYSIAL CELLS IN AGING LOBUND-WISTAR RATS IN TISSUE CULTURE: AN ULTRASTRUCTURAL STUDY

Sylvia L. Asa, M.D., Kalman Kovacs, M.D., Ph.D., Blair M. Gerrie, M.Sc., Robin E. Baird, and Gezina Ilse, R.T.
Department of Pathology, St. Michael's Hospital, University of Toronto, Toronto, Ontario, Canada.

INTRODUCTION

Pituitary adenomas are known to occur spontaneously in old rats of several strains (Wolfe et al., 1938; Saxton and Graham, 1944; Wolfe and Wright, 1947; Kim et al., 1960; Griesbach, 1967; Ito et al., 1972; Kovacs et al., 1977; Berkvens et al., 1980; McComb et al., 1981; Trouillas et al., 1982; McComb et al., 1984, 1985a). The majority of the tumors studied have been prolactin-producing adenomas composed of lactotrophs and associated with hyperprolactinemia (Kim et al., 1960; Kovacs et al., 1977; McComb et al. 1981; Trouillas et al., 1982; McComb et al., 1984). Tumor types have been described which produce the other adenohypophysial hormones and may be mono- or plurihormonal (Berkvens et al., 1980; McComb et al., 1981, 1984, 1985a, b).

The spontaneous occurrence of pituitary adenomas in these animals offers an experimental model for the pituitary tumors of humans. The pathogenetic factors underlying these lesions are unclear. It has been suggested that variation in the hormonal milieu may be important in the regulation of adenohypophysial cell proliferation both in humans (Asa and Kovacs, 1984) and in experimental animals (Furth et al., 1973, 1976). The study of lactotroph adenomas in rats has shown that a decrease in hypothalamic inhibition of prolactin release can result in hyperprolactinemia and lactotroph proliferation (Meites, 1982; Sarkar et al., 1982, 1983 a, b). Lactotroph adenomas were more common in rats who had decreased hypothalamic concentration of the prolactin-inhibiting factor dopamine or who had defective dopamine release or transport to the adenohypophysis (Sarkar et al., 1982). Some investigators have shown that in aging rats pituitary neovascularization with systemic non-portal vessels exposes the lactotrophs to dopamine-poor blood and may play a role in lactotroph hyperplasia and neoplasia (Elias and Weiner, 1984, 1987; Schechter et al., 1987). Other investigators have documented that prolonged administration of estrogen causes prolactin cell hyperplasia and neoplasia in some strains of aging rats (El Etreby and Günzel, 1974; Lloyd, 1983).

The use of a controlled environment as in tissue culture, provides an opportunity to study structure-function correlations in various states of hormonal control. Therefore the pituitaries of aging Lobund-Wistar rats were studied in vitro and the morphologic features of cultured pituitary tissues were compared with those obtained at the time of sacrifice of the animal.

MATERIALS AND METHODS

Male Lobund-Wistar rats were kept at the Lobund Laboratory, University of Notre Dame. The details of the design of the project, the conditions of animal environments and the growth and survival characteristics of the animals are described elsewhere in this book. Animals at 6, 18, and 30 months were killed by exsanguination under halothane anesthesia. Several animals between the ages of 30 and 43 months were also studied.

Pituitaries were divided for morphologic and tissue culture studies. The methods of fixation and embedding for morphologic examination are described in the previous chapter.

For tissue culture, sterile tissues were placed aseptically in culture medium containing antibiotics, kept on ice and sent by overnight courier from the Lobund Laboratory to St. Michael's Hospital, Toronto, Ontario, Canada. The tissue was immediately washed two or three times in sterile medium (CMRL-1969, Connaught, Willowdale, Ontario, Canada), then minced into small pieces. Cell dispersion was performed by gentle agitation in a spinner flask (Belco Glass Inc., Vineland, NJ) containing an enzyme solution of collagenase (1 mg/ml CMRL-1969) incubated at 37°C for 30-45 min. with periodic trituration. The yield and viability of cells was estimated with a hemocytometer and light microscope using the trypan blue exclusion technique. Dispersed cells were harvested by centrifugation and suspended in medium supplemented with 10% fetal calf serum (Grand Island Biological Co., Grand Island, NY). Multiwell culture dishes (Linbro, Flow Laboratories Inc., McLean, VA) were coated with a thin layer of purified collagen (Vitrogen 100, Flow Laboratories) and allowed to air dry at room temperature. Dispersed cells were plated onto collagen-coated incubation chambers 10^4 cells/500 µl/well. Cultures were maintained in a CO_2 incubator (Napco, Portland, OR) at 37°C in a humidified atmosphere of 95% air - 5% CO_2. Cells were allowed to attach for 2-3 days. Subsequently, media were collected every 24-72 hours and stored in polyethylene vials at -20°C for future radioimmunoassay analysis.

At the termination of cultures (7-10 days), monolayers were treated with 10 mg/ml purified trypsin (Worthington, Freehold, NJ) for 10 min., harvested by centrifugation, washed in medium and centrifuged into pellets. Pellets were fixed in 2.5% glutaraldehyde in Sorensen's buffer, postfixed in 1% osmium tetroxide in Millonig's buffer, dehydrated in graded ethanols and embedded in epoxy resin. Ultrathin sections were stained with uranyl acetate and lead citrate and examined with a Philips 410-LS electron microscope (Philips Electronic Instruments, Mahway, NJ).

RESULTS

The pituitaries of 97 male Lobund-Wistar rats were studied *in vitro*. The environmental conditions and ages of the animals are shown in Table 1. Cell viability averaged 81%.

After tissue culture, adenohypophysial cells of all five types were identified. The cells had features similar to those of the gland *in vivo* with two major exceptions; the number and size of lysosomes were strikingly increased and the number of secretory granules was reduced.

Table 1. Characteristics of Experimental Animals Whose Pituitaries were Studied In Vitro

Environmental Conditions	Ages (months)			
	6	18	30	>30
Conventional Fullfed	6	--	10	4
Conventional Restricted	9	8	7	6
Germfree Fullfed	8	--	7	5
Germfree Restricted	9	--	9	9

Somatotrophs in culture had short rough endoplasmic reticulum (RER) profiles, small Golgi regions and sparse electron dense secretory granules. Lactotrophs were present in moderate numbers; there was no evidence of hyperplasia nor were adenomatous lactotrophs identified. The cells were large with well developed RER, prominent Golgi regions harboring forming secretory granules and occasional large secretory granules. The lactotrophs in culture appeared less active than those in vivo. Corticotrophs resembled those in the native gland but contained more numerous lysosomes. Thyrotrophs in the nontumorous gland in vivo contained numerous small secretory granules filling the cytoplasm; in contrast, cultured thyrotrophs maintained well developed RER and Golgi complexes but stored few small granules, most prominently at the cell periphery.

The most striking alterations in cell morphology were seen in gonadotrophs. In the pituitaries of 6 and 18 months old rats, gonadotrophs were large, round cells found frequently in association with lactotrophs. They contained abundant, well developed vesicular RER and large paranuclear Golgi areas. Two populations of secretory granules were identified; small dense granules measuring about 200 nm in diameter were intermixed with large, moderately dense granules of up to 1000 nm diameter. In rats 30 months of age or older, some gonadotrophs appeared to show features of castration cells. This population was striking in the fullfed group but was inconspicuous in the animals maintained under dietary restriction. The castration cells were hypertrophied and their cytoplasm was almost filled with dilated, enlarged endoplasmic reticulum vesicles containing a flocculent, moderately electron dense material. The enlarged ring-like Golgi complex contained forming secretory granules and occasional secretory granules were found scattered in the cytoplasm, however, the granule number was significantly lower than in resting gonadotrophs of the younger or dietary restricted animals. In tissue culture, the gonadotrophs identified had well developed RER profiles, large vesicular paranuclear Golgi regions and intermediate numbers of secretory granules ranging from 200 to 1000 nm in diameter. There was no variability in the morphology of cells from the different groups of animals and no castration cells were found. Gonadotrophs were numerous in cultured tissues, however, no adenomatous gonadotrophs were identified with certainty.

A population of cells with accumulations of large numbers of mitochondria was identified in cultures of the pituitaries of some old fullfed rats. These oncocytes had few granules whose morphology resembled those of gonadotrophs. Cells with minimal oncocytic changes were found in a few gonadotrophs in vivo, but no true oncocytes were identified in those tissues. In contrast, a few gonadotroph adenomas in vivo were composed of cells with poorly developed cytoplasmic organelles and few cytoplasmic granules but no increase in mitochondrial number, cells with those features were infrequent in cultured tissues.

DISCUSSION

The factors underlying the development of pituitary adenomas are unknown. In humans, some types of adenomas have been associated with longstanding target organ failure. Thyrotroph adenomas have been reported in patients with untreated hypothyroidism (Katz et al., 1980; Scheithauer et al., 1985) and have been observed in rats rendered hypothyroid (Furth et al., 1973). Gonadotroph adenomas have occurred in hypogonad patients (Kovacs et al., 1980; Nicolis et al., 1988) and basophil adenomas have been identified in gonadectomized rats (Griesbach and Purves, 1960). Although prolonged untreated hypocorticism is incompatible with life, patients who have died with chronic Addison's disease have had nodular and diffuse corticotroph hyperplasia (Scheithauer et al., 1983) and patients with Addison's disease who have been treated with corticosteroid substitution have had evidence of ACTH secreting adenomas (Krautli et al., 1982).

The mechanism of tumor growth enhancement may involve release from feedback inhibition of target organ hormones; alternatively it may be attributed to chronic excessive stimulation by hypothalamic regulatory peptides. In experimental situations, growth hormone-releasing hormone (GRH) and corticotropin-releasing hormone (CRH) have been shown to enhance somatotroph and corticotroph proliferation respectively (Billestrup et al., 1986; Gertz et al., 1987) and patients with excessive production of GRH and CRH by tumors are known to have somatotroph and corticotroph hyperplasia (Sano et al., 1988; Asa et al., 1984a; Carey et al., 1984). While these studies suggest that the peptides stimulate cell proliferation, they do not implicate them as a cause of tumorigenesis. The multistep theory of carcinogenesis (Farber, 1981) would provide a mechanism whereby stimulatory peptides may act as promotors of tumor growth. The association of pituitary somatotroph adenomas with sellar or parasellar GRH-producing gangliocytomas supports this possibility (Asa et al., 1984b; Sano et al., 1988).

The occurrence of pituitary gonadotroph adenomas in aging Lobund-Wistar rats provides a unique model to study the pathogenesis of spontaneous gonadotroph tumors and to clarify the effects of decreased target organ feedback and/or excessive hypothalamic stimulation in the etiology and growth of these adenohypophysial neoplasms. Our study has shown that adenohypophysial cells of Lobund-Wistar rats persist in tissue culture and maintain characteristic morphology. Decreased numbers of secretory granules and reduced development of RER in cultured cells suggest that those organelles are dependent on the hormonal milieu of cells in vivo. These results are similar to those we have found in cultured human gonadotroph adenomas (Asa et al., 1988). The similarities of the human and rat gonadotroph adenomas in vitro also make this animal model an ideal one for the study of human gonadotroph adenomas.

In this study, gonadectomy cells were found in pituitaries of aging fullfed rats but were not conspicuous in those of rats maintained under dietary restriction. The presence of gonadectomy cells suggests that target organ insufficiency may play a role in the genesis or growth of

the pituitary adenomas in these animals. The effects of
dietary restriction on the development of gonadectomy cells
remain to be clarified. The lack of gonadectomy changes in
vitro confirms that this change is reversible and reflects
the hormonal milieu of the cells in vivo.

Several gonadotroph adenomas in these experimental
animals were composed of poorly differentiated cells with
scant cytoplasmic organelles. These adenomas resembled
null cell adenomas of the human pituitary (Kovacs and
Horvath, 1986), however, they contained immunoreactive
gonadotropins. These data support the evidence for
gonadotropic differentiation of human null cell adenomas
provided by our tissue culture studies (Asa et al., 1986).
Moreover, oncocytic change was found in cultures of these
animal pituitaries whereas oncocytes were not seen in vivo.
It has been suggested that oncocytic change occurs most
frequently in null cell adenomas (Asa et al., 1986; Kovacs
and Horvath, 1986), however, the factors underlying this
transformation are not known. The appearance of oncocytes
in vitro suggests that this phenomenon may reflect reaction
to the cellular environment; identification of oncocytic
change in gonadotrophs of these experimental animals sheds
light on the cytogenesis of oncocytic adenomas of humans.

ACKNOWLEDGEMENT

This work was supported in part by The St. Michael's
Hospital Research Society and Grant MT-6349 of the Medical
Research Council. The authors wish to thank Dr. Eva
Horvath for her help in evaluating the electron microscopic
results and Mrs. Hazel Douglas for secretarial work.

REFERENCES

Asa SL, Kovacs K (1984). Development and proliferation of
  adenohypophysial cells. In: Evolution and Tumour
  Pathology of the Neuroendocrine System. Ed. by Falkmer
  S, Hakanson R, Sundler F, Elsevier, Amsterdam, p. 399.
Asa SL, Kovacs K, Tindall GT, Barrow DL, Horvath E, Vecsei
  P (1984a). Cushing's disease associated with an
  intrasellar gangliocytoma producing corticotropin-
  releasing factor. Ann Intern Med 101:789.
Asa SL, Scheithauer BW, Bilbao JM, Horvath E, Ryan N,

Kovacs K, Randall RV, Laws ER Jr, Singer W, Linfoot JA, Thorner MO, Vale W (1984b). A case for hypothalamic acromegaly: a clinicopathological study of six patients with hypothalamic gangliocytomas producing growth hormone releasing factor. J Clin Endocrinol Metab 59:796.

Asa SL, Gerrie BM, Singer W, Horvath E, Kovacs K, Smyth HS (1986). Gonadotropin secretion in vitro by human pituitary null cell adenomas and oncocytomas. J Clin Endocrinol Metab 62:1011.

Asa SL, Gerrie BM, Kovacs K, Horvath E, Singer W, Killinger DW, Smyth HS (1988). Structure-function correlations of human pituitary gonadotroph adenomas in vitro. Lab Invest 58:403.

Billestrup N, Swanson LW, Vale W (1986). Growth hormone-releasing factor stimulates proliferation of somatotrophs in vitro. Proc Natl Acad Sci USA 83:6854.

Berkvens JM, Van Nesselrooy JH, Kroes R (1980). Spontaneous tumors in the pituitary gland of old Wistar rats. A morphologic and immunocytochemical study. J Pathol 130:179.

Carey RM, Varma SK, Drake CR Jr, Thorner MO, Kovacs K, Rivier J, Vale W (1984). Ectopic secretion of corticotropin-releasing factor as a cause of Cushing's syndrome: a clinical, morphological and biochemical study. New Engl J Med 311:13.

El Etreby MF, Gunzel P (1974). Sex hormones - effects on prolactin cells. Acta Endocrinol 76 (Suppl 189):166.

Elias KA, Weiner RI (1984). Direct arterial vascularization of estrogen-induced prolactin secreting anterior pituitary tumors. Proc Natl Acad Sci USA 81:4549.

Elias KA, Weiner RI (1987). Inhibition of estrogen-induced anterior pituitary enlargement and arteriogenesis by bromocriptine in Fischer 344 rats. Endocrinology 120:617.

Farber E (1981). Chemical carcinogenesis. New Engl J Med 305:1379.

Furth J, Ueda G, Clifton KH (1973). The pathophysiology of pituitaries and their tumors: Methodological advances. In: Methods in Cancer Research. Ed. by Bush H, Academic Press, New York, p. 201.

Furth J, Nakane P, Pasteels JL (1976). Tumors of the pituitary gland. In: Pathology of Tumors in Laboratory Animals. Vol 1. Ed. by Turosov US. IARC Scientific Publication, Lyon, France, p. 201.

Gertz BJ, Conteras LN, McComb DJ, Kovacs K, Tyrrell JB, Dallman MF (1987). Chronic administration of corticotropin-releasing factor increases pituitary corticotroph number. Endocrinology 120:381.

Griesbach WE (1967). Basophil adenomata in the pituitary glands of 2-year-old male Long-Evans rats. Cancer Res 27:1813.

Griesbach WE, Purves HD (1960). Basophil adenomata in the rat hypophysis after gonadectomy. Br J Cancer 14:49.

Ito A, Moy P, Kaunitz H, Kortwright K, Clarke S, Furth J, Meites J (1972). Incidence and character of spontaneous pituitary tumors in strain CR and W/Fu male rats. J Natl Cancer Inst 49:701.

Katz MS, Gregerman RI, Horvath E, Kovacs K, Ezrin C (1980). Thyrotroph cell adenoma of the human pituitary gland associated with primary hypothyroidism: Clinical and morphological features. Acta Endocrinol 95:41.

Kim U, Clifton KH, Furth J (1960). A highly inbred line of Wistar rats yielding spontaneous mammosomatotropic pituitary and other tumors. J Natl Cancer Inst 24:1031.

Kovacs K, Horvath E (1986). Tumors of the Pituitary Gland. Atlas of Tumor Pathology, Second Series, Fascicle 21, Armed Forces Institute of Pathology, Washington, D.C.

Kovacs K, Horvath E, Ilse RG, Ezrin C, Ilse D (1977). Spontaneous pituitary adenomas in aging rats. A light microscopic, immunocytochemical and fine structural study. Beitr Pathol 161:1.

Kovacs K, Horvath E, Rewcastle NB, Ezrin C (1980). Gonadotroph cell adenoma of the pituitary in a woman with long-standing hypogonadism. Arch Gynecol 229:57.

Krautli B, Müller J, Landolt AM, von Schulthess F (1982). ACTH-producing pituitary adenomas in Addison's disease: two cases treated by transsphenoidal microsurgery. Acta Endocrinol 99:357.

Lloyd RV (1983). Estrogen-induced hyperplasia and neoplasia in the rat anterior pituitary gland. An immunohistochemical study. Am J Pathol 113:198.

McComb DJ, Hellmann P, Kovacs K, Scott D, Evans WS, Burdman JA, Thorner MO (1985a). Spontaneous sparsely granulated prolactin-producing pituitary adenomas in aging rats: a prospective study of the effect of bromocriptine. Neuroendocrinology 41:201.

McComb DJ, Kovacs K, Beri J, Zak F (1984). Pituitary adenomas in old Sprague-Dawley rats: A histologic, ultrastructural, and immunocytochemical study. J Natl Cancer Inst 73:1143.

McComb DJ, Kovacs K, Beri J, Zak F, Milligan JV, Shin SH (1985b). Pituitary gonadotroph adenomas in old Sprague-Dawley rats. J Submicrosc Cytol 17:517.

McComb DJ, Ryan N, Horvath E, Kovacs K, Nagy E, Berczi I,

Domokos I, Laszlo FA (1981). Five different adenomas derived from the rat adenohypophysis: immunocytochemical and ultrastructural study. J Natl Cancer Inst 66:1103.

Meites J (1982). Changes in neuroendocrine control of anterior pituitary function during aging. Neuroendocrinology 34:151.

Nicolis G, Shimshi M, Allen C, Halmi NS, Kourides IA (1988). Gonadotropin-producing pituitary adenoma in a man with long-standing primary hypogonadism. J Clin Endocrinol Metab 66:237.

Sano T, Asa SL, Kovacs K (1988). Growth hormone-releasing hormone (GRH)-producing tumors: clinical, biochemical and morphologic manifestations. Endocrine Rev 9:in press.

Sarkar DK, Gottschall PE, Meites J (1982). Damage to hypothalamic dopaminergic neurons is associated with development of prolactin-secreting pituitary tumors. Science 218:684.

Sarkar DK, Gottschall PE, Meites J (1983a). Relation of the neuroendocrine system to development of prolactin-secreting pituitary tumors. In: Neuroendocrinology of aging. Ed. by Meites J, Plenum, New York, pp. 353.

Sarkar DK, Miki N, Meites J (1983b). Failure of prolactin short loop feedback mechanism to operate in old as compared to young female rats. Endocrinology 113:1452.

Saxton JA, Graham JB (1944). Chromophobe adenoma-like lesions of the rat hypophysis. Frequency of the spontaneous lesions and characteristics of growth of homologous intraocular transplants. Cancer Res 4:168.

Schechter J, Ahmad N, Elias K, Weiner R (1987). Estrogen-induced tumors: changes in the vasculature in two strains of rats. Am J Anat 179:315.

Scheithauer BW, Kovacs K, Randall RV (1983). The pituitary gland in untreated Addison's disease. Arch Pathol Lab Med 107:484.

Scheithauer BW, Kovacs K, Randall RV, Ryan N (1985). Pituitary gland in hypothyroidism: Histologic and immunocytologic study. Arch Pathol Lab Med 109:499.

Trouillas J, Girod C, Claustrat B, Curé M, Dubois MP (1982). Spontaneous pituitary tumors in the Wistar/Furth/Ico rat strain. Am J Pathol 109:57.

Wolfe JM, Bryan WR, Wright AW (1938). Histologic observations on the anterior pituitaries of old rats with particular reference to the spontaneous appearance of pituitary adenomata. Am J Cancer 34:352.

Wolfe JM, Wright AW (1947). Cytology of spontaneous adenomas in the pituitary gland of the rat. Cancer Res 7:759.

DIETARY RESTRICTION AND HYPOTHALAMIC METABOLISM: A MODEL OF AGING AND FOOD INTAKE CONTROL

Roy J. Martin

Department of Foods and Nutrition, University of Georgia, Athens, GA 30602

INTRODUCTION

Restricted feeding of animals has been used as a model for food intake control and as a model for aging research. In studies of aging, it is clear that food restriction is the most powerful nutritional manipulation of the aging process (McKay et al., 1935; Young, 1979; Masoro et al., 1980). In studies of food intake control, food restriction is an extremely strong stimulus for the initiation of feeding behavior (Harris and Martin, 1984a). Furthermore, it is known that obesity or overfeeding results in a decrease of life expectancy (Van Itallie, 1985). Therefore, there are very likely some overlapping mechanisms for both the regulation of the aging process and the regulation of energy balance. The metabolic changes in the CNS which may be involved in energy balance may also provide the neurochemical signals which enhance longevity.

The following is a brief review of some of our research on the metabolic adjustments made to restriction and overfeeding. It has been shown that restricted feeding caused a loss of body fat, below the "set point" and resulted in the animals being in a state of hunger. Conversely, overfeeding caused an increase in body fat above the set point and resulted in the animals being in a state of satiety (Fig. 1). If the restricted fed animal is allowed to eat, it overate until it regained its original body weight. On the other hand, the over fed (force fed) animal underate until its original body weight was achieved. The regulation of energy balance under these conditions is thought to be controlled by endogenous signals that reflect the energy status of the animal (Harris et al., 1984b; Harris et al., 1986).

We have proposed that there were certain areas of the brain that recognized altered energy balance. It was also proposed that this

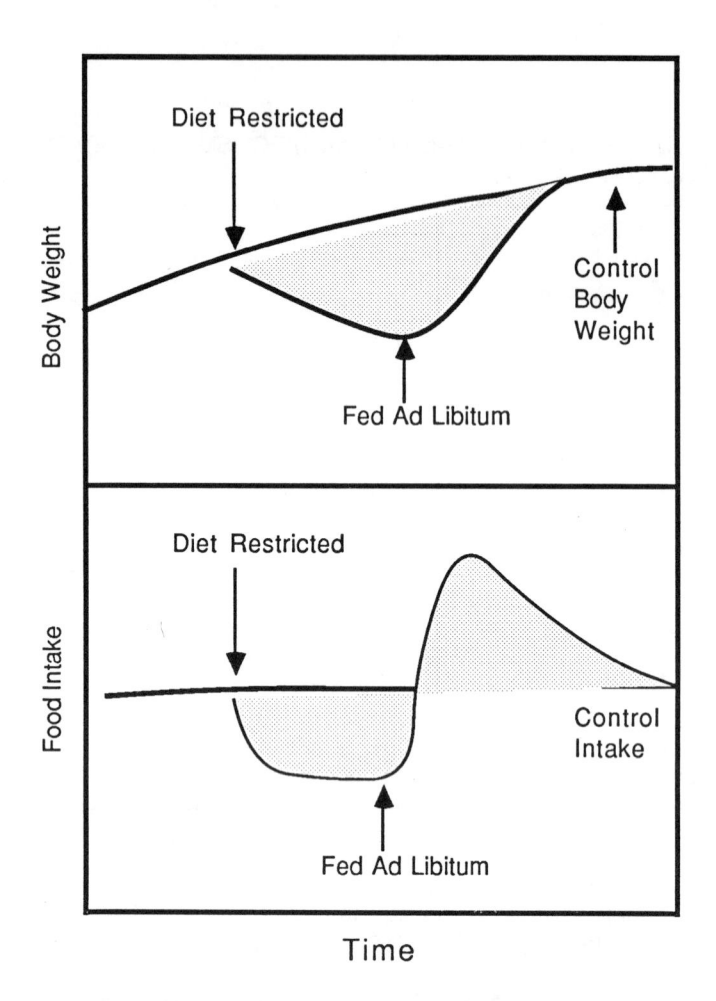

Figure 1. Diet Restricted Animal Model for the Study of the Control of Feeding Behavior.

recognition was achieved by a change in brain metabolic pathways which subsequently led to an alteration in neurochemical signals (Kasser et al., 1985a). There are several questions which when answered supported this line of thinking.

What Area of the Brain Is Predominately Involved in Energy Balance Regulation ?

The hypothalamus has been known to be involved in the regulation of feeding and metabolic status (Anad and Brobeck, 1951; Luiten et al 1986, Bernardis and Bellinger 1987). Lesions in the ventromedial hypothalamus caused an increase in feeding and body weight, whereas lesions in the lateral hypothalamus resulted in a decrease in feeding. Indeed there are other areas involved in the control of feeding; however, when lesioned they did not produce as dramatic a change in metabolism and body weight as when the ventromedial or lateral hypothalamus was lesioned.

Why Look at Hypothalamic Metabolism During Restricted Feeding?

Nicolaidis (1978) proposed that the drive to eat was influenced by metabolism and that there were cells in the hypothalamus that acted as detectors of peripheral metabolism. He showed that lipid content of the hypothalamus was reflective of body fat content (Nicolaidis et al.,1974). Changes found in hypothalamic metabolism of hungry and satiated rats further supported Nicholaidis' proposal (Panksepp, 1974). Oxygen consumption of the lateral hypothalamus when expressed relative to the oxygen consumption of the medial hypothalamus was elevated in the feeding state. Oomura (1981) provided more direct evidence that simple energy substrates such as glucose and fatty acids could influence the activity of selective neurons in the lateral and ventromedial hypothalamus. Taken together these observations point to a role for brain metabolism in the control of food intake.

Does the Pattern of Brain Uptake of Metabolites Reflect Energy Balance State?

During starvation and energy restriction brain uptake of ketone bodies and fatty acids are increased. We have shown that fatty acid and glucose uptake into hypothalamic areas were influenced by energy restriction (Kasser and Martin, 1985). The brain uptake index (BUI) for fatty acids was increased 3 fold in the restricted animals while the BUI for glucose was decreased 30% in the same areas. The pattern of brain uptake of metabolites does reflect energy balance state. These observations supported the contention that brain uptake of metabolites reflected energy balance states.

Does the Metabolic Pattern of the Hypothalamus Reflect Changes Seen in Peripheral Metabolism?

We have characterized fatty acid and glucose metabolism in three brain sites associated with feeding (Kasser et al, 1985a, 1985b). Rats were tube fed a liquid diet at three levels of intake (50, 100 and 150% of normal intake) for 7 days. The body fat content was influenced by the level of feeding as expected. Restricted feeding resulted in an increase in lateral hypothalamic fatty acid oxidation and a decrease in fatty acid synthesis in the medial hypothalamus. The overall utilization of glucose was not affected by energy status, however the individual pathways of glucose utilization were affected. Glucose metabolized by the GABA shunt and pentose shunt pathway were altered in selective areas of the hypothalamus (Table 1). These observations demonstrated that within selective brain sites, specific pathways for glucose oxidation were affected by energy intake. For these metabolic changes to be used to assess and respond to changes in peripheral energy status, it was necessary to show that these metabolic patterns be maintained during the recovery of body weight after over or under feeding. Thus far, fatty acid oxidation in the lateral hypothalamus remains elevated until body weight is regained after refeeding the restricted fed animal. Thus, the metabolic pattern of the hypothalamus does reflect changes seen in peripheral metabolism.

TABLE 1. Specific Hypothalamic Metabolic Adjustments to Restricted Feeding

Metabolic Parameter	Response to Restricted Feeding
Lateral hypothalamus	
Fatty Acid Oxidation	Increased
GABA Shunt Pathway	Decreased
Medial Hypothalmus	
Fatty Acid Synthesis	Decreased
Pentose Shunt Pathway	Decreased
Glutamate Decarboxylase	Increased
Brain Stem Areas	
Pentose Shunt Pathway	Decreased

How Does Activity of a Metabolic Pathway Become Translated to a Neurochemical Signal?

This question can only be answered in a theoretical manner for there are no direct answers at this time. The most direct manner in which changes in glucose metabolism can be translated to a neurochemical signal may be through changes in the rate of glucose metabolism through the GABA shunt pathway. As more glucose is metabolized through this pathway, more of the neurotransmitter GABA can be produced. GABA is an inhibitory neurotransmitter. In the ventromedial hypothalamus GABA is associated with an increase in feeding (Pansepp and Meeker, 1980). We have found that the rate limiting enzyme in this pathway, glutamate decarboxylase, was elevated in the ventromedial hypothalamus of the restricted fed animal (Beverly and Martin, 1988).

Glucose metabolism through the pentose phosphate pathway was also influenced by restricted feeding. The pentose phosphate pathway was associated with the production of reducing equivalents for biogenesis of amines (Gaitonde et al., 1983). Norepinephrine turnover may be influenced by restricted feeding through alterations in the metabolism of glucose through this pathway. It has recently been demonstrated that production of GABA in the brain could also proceed through the pentose phosphate pathway (Gaitonde et al., 1987). Inhibition of the pentose phosphate pathway in rat brain by 6-aminonicotinamide resulted in a 57% decrease in the incorporation of glucose carbon into GABA.

Another mechanism by which neuronal metabolic activity may influence brain function is through the alteration of receptors. In the periphery, cellular metabolism is thought to influence hormone response status by either influencing the recycling of receptors to the plasma membrane (Fig. 2) or through the influence on post receptor mechanisms involving protein kinases (Benovic et al., 1986). Metabolism of neurons may alter the response to neuropeptides in a similar manner.

CONCLUSIONS

The pattern of metabolism of the brain is influenced by dietary restriction. It is not proven that this metabolic information can serve as a signal of energy status, however there are some mechanisms by which metabolic information can be translated to a neurochemical signal. The influence of aging on neurotransmitter function has been well documented (Simpkins and Millard, 1987). Dopamine and norepinephrine turnover is reduced in the hypothalamus during aging. Restricted feeding, which increases life span in rats, leads to several changes in brain metabolism. These changes in brain metabolism may influence brain function through an alteration in neurotransmitter biogenesis or through alteration of neuropeptide function.

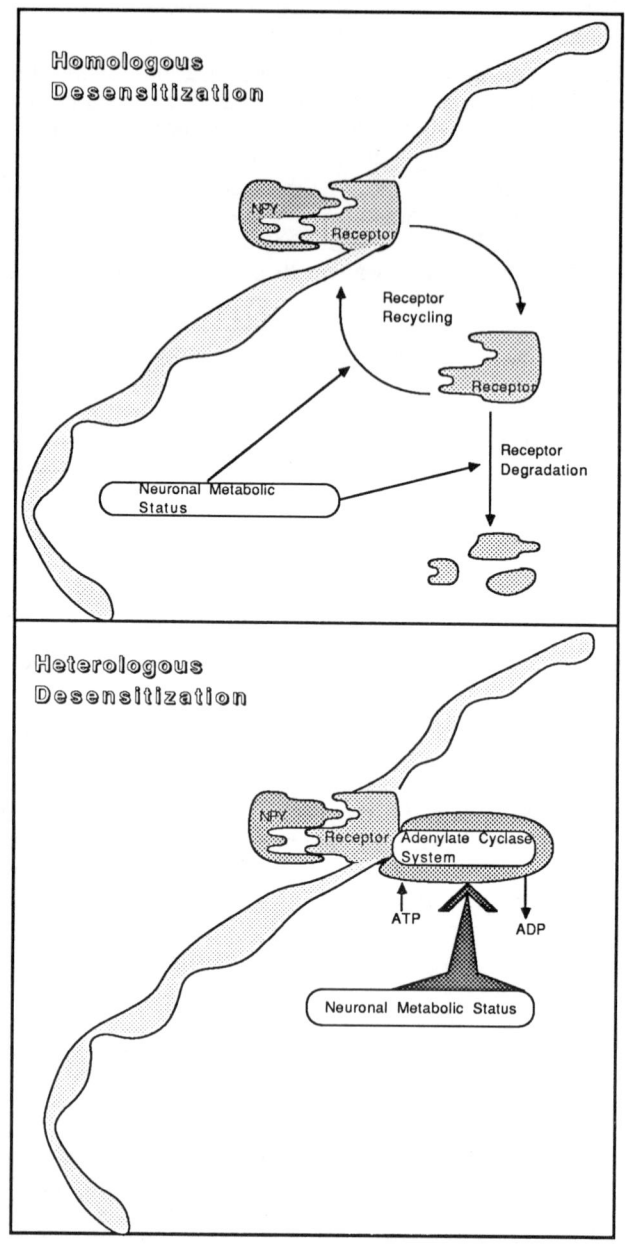

Figure 2. Proposed Mechanisms by Which Neuronal Metabolism Influences Responsiveness to Neuromodulator. NPY= Neuropeptide Y.

RERERENCES

Anand BK, Brobeck JR (1951). Hypothalamic control of food intake in rats and cats. Yale J Biol Med 24:123-129.

Benovic JL, Strasser RH, Caron MG, Lefkowitz (1986). Adrengic receptor kinase: Identification of a novel protein kinase that phosphorylates the agonist-occupied form of the receptor. Proc Natl Acad Sci 83:2797-2801.

Bernardis LL, Bellinger LL(1987). The dorsomedial hypothalamic nucleus revisited:1986 update. Brain Res Rev 12:321-381.

Beverly JL, Martin RJ (1988). Increased GABA shunt activity in the VMN of three hyperphagic rat models. (Submitted)

Gaitonde MK, Jones J, Evans G (1987). Metabolism of glucose into glutamate via the hexose monophosphate shunt and its inhibition by 6-aminonicotinamide in rat brain in vivo. Proc R Soc Lond B 231:71-90.

Gaitonde MK, Evison E, Evans GM (1983). The rate of utilization of glucose via hexosemonophosphate shunt in brain. J Neurochem 41:1253-1260.

Harris RBS, Kasser TR, Martin RJ (1986). Dynamics of recovery of body composition after overfeeding , food restriction or starvation of mature female rats. J Nutr 116:2536-2546.

Harris RBS, Martin RJ (1984a). Lipostatic theory of energy balance: Concepts and signals. Nutr Behav 1:253-275.

Harris RBS, Martin RJ (1984b). Recovery of body weight from below "set point" in mature female rats. J Nutr 114:1143-1150.

Kasser TR, Harris RBS, Martin RJ (1985a). Level of satiety: fatty acid and glucose metabolism in three brain sites associated with feeding. Am J Physiol 248: R447-R452.

Kasser TR, Harris RBS, Martin RJ (1985b). Level of satiety: GABA and pentose shunt activities in three brain sites associated with feeding. Am J Physiol 248: R453-R458.

Kasser TR, Martin RJ (1986). Induction of ventrolateral hypothalamic fatty acid oxidation in diabetic rats. Physiol Behav 36:385-388.

Kasser TR, Deutch A, Martin RJ (1985). Uptake and utilization of metabolites in specific brain sites relative to feeding behavior. Physiol Behav 36:1161-1165.

Luiten PGM, ter Horst G J, Steffens AB (1987). The hypothalamus, intrinsic connections and outflow pathways to the endocrine system in relation to the control of feeding and metabolism. Prog Neurobiol 28:1-54.

Masoro SJ, Yu BP, Bertrand HA, Lynd FT (1980). Nutritional probe of the aging process. Fed Proc 39:3178-3182.

McCay CM, Crowell MF, Maynard LA (1935). The effect of retarded growth upon the life span and upon the ultimate body size. J Nutr 10:63-79.

Nicolaidis S (1978). Role des reflexes anticipateurs orovegetatifs dans la regulation hydrominerale et energetic. J Physiol Paris 74:1-19.

Nicolaidis S, Petit M, Polonowski J (1974). Etude du rapport entra la regulation de la masse adipeuse corporelle et la composition lipidique de ses centres regulateurs. C R Natl Acad Sc , Paris 278:1393-1396.

Oomura Y (1981). Chemosensitive neuron in the hypothalamus related to food intake and behavior. Jpn. J. Pharmacol. 30: 1-12.

Panksepp J (1974). Hypothalamic regulation of energy balance and feeding behavior. Fed Proc 33:1150-1153.

Pankseep J, Meeker RB (1980). The role of GABA in the ventromedialhypothalamic regulation of food intake. Brain Res Bull 5:543-460.

Simkins JW, Millard WJ (1987). Influence of age on neurotransmitter function. Endocrin Aging 16:893-918.

Van Itallie TB (1985). Health implications of overweight and obesity in the United States. Ann Intern Med 103:983-988.

Young VR (1979). Diet as a modulator of aging and longevity. Fed Proc 38:1994-2000.

GASTROINTESTINAL PHYSIOLOGY

EFFECT OF DIETARY RESTRICTION ON GASTROINTESTINAL CELL GROWTH

Peter R. Holt

Division of Gastroenterology, Department of Medicine, St. Luke's Hospital Center and College of Physicians & Surgeons of Columbia University, New York, NY 10025

INTRODUCTION

The question for which we have virtually no answers is "Does the gastrointestinal tract age and, if it does, what are the disorders that follow?" Clearly, the gastrointestinal tract differs intrinsically from the organs that usually have been investigated as a function of age; for example, the liver and brain. The gastrointestinal tract, from mouth to anus, consists of replicating tissues and replication in much of the intestine, particularly the small intestine, is extraordinarily rapid. It has been estimated that small intestinal epithelial cells "turn over" in two to three days in the rat and in no more than about five or so days in man and that the colon, stomach and esophagus also show extremely rapid replication. One would therefore anticipate that the manifestation of aging in such tissues intrinsically would differ to a considerable extent from age-related changes in organs containing predominantly post-mitotic cells such as the liver and brain. Certainly it is of interest to compare and contrast age-related changes in predominantly post-mitotic and predominantly replicating organs for such comparisons may shed light on intracellular events that are generalizable to all cells and events that may be specific to either post-mitotic or replicating cells.

To understand the basic physiology that characterizes cells of the gastrointestinal tract, it is necessary to describe the anatomic and physiologic components of cell production, cell differentiation and cell migration to appreciate how these components might be altered as a function of age. Cell division occurs in the crypt in a zone of proliferation which overall is reproducibly definable for each gastrointestinal organ. Cells then migrate up the crypt and, in the small intestine, pass into a differentiating zone around the area of the crypt-villus junction to attain fully differentiated features in the area of the mid-villus. Such differentiated features in the small intestine are characterized by maximal activity of a series of brush border enzymes which include the disaccharidases, sucrase, lactase and maltase, as well as alkaline phosphatase and other enzymes which hydrolyze luminal nutrients at the microvillus border. Epithelial cells also change histologically between the crypt and villus so that differentiated villus cells contain a well-organized microvillus brush border and endoplasmic reticulum. Such cells have functional characteristics that are directed toward maximal and effective nutrient transport. Similar differentiation also occurs in colon and stomach although the precise regions for such differentiation may differ from those found in the small intestine (Lipkin, 1981).

Normally, in the steady state, cell production in the replicating intestine is determined by the rate of cell migration and of cell loss. Thus, for example, cell proliferation may be increased when cell loss is exaggerated following injury to the intestine. Cell production is "turned down" when less is needed; for example, during starvation. If a small intestinal segment is resected and removed, the remaining small intestine, as well as the colon, will proliferate. Similarly, when feeding follows a period of starvation, a burst of proliferation occurs in small and large intestine. Such proliferation is characterized by an increase in the number of crypt cells incorporating ^{3}H-thymidine into DNA, an increase in the specific activity or total activity of enzymes that are associated with proliferation, such as thymidine kinase, and a reduction in cell cycling time. This is followed by an increase in the rate of migration of the newly formed cells from the crypt up the villus.

Previously, that is before 1980, the few studies that had been performed upon the effect of aging on the gastrointestinal tract implied that several functional characteristics were impaired. Cell production rate in the small intestine was said to be lower in the studies of Lesher and co-workers (Lesher et al, 1962; Fry et al, 1961). In the proximal intestine, cell cycling time was found to be prolonged (Thrasher et al, 1967). Cell migration, when studied in the mouse, appeared to be slower and several investigators had suggested that the structure of the villus compartment of the small intestine was altered. A shortening of villi were described in aging animals (Hohn et al, 1978) and in man (Webster and Leeming, 1975). These observations were not uniform since other investigators later have found no changes in villi of aging animals (Ecknauer, 1982) or in man (Corazza et al, 1986). In addition, previously several functional components also were said to be altered. Characteristically, impaired carbohydrate absorption had been described (Phillips and Gilder, 1940) and xylose absorption also has been found to be defective (Kendall, 1970; Webster and Leeming, 1975).

The studies in our own laboratory originally were designed to test a hypothesis that was based upon some of these previous observations. The hypothesis that we tested was that if structural changes occur as a function of aging in the small intestine and if functional changes were present, then both of these may be related to an alteration of cellular proliferation, i.e., a slowdown of the proliferative response. Our studies, which started in 1980, were designed to test this hypothesis and used as an experimental animal model the _ad lib_ fed male Fischer 344 rat. Animals at two extremes of life initially were studied--as young control animals, Fischer rats 4-5 mo of age and, as aging animals, Fischer rats that were 27 mo of age. Animals at these ages were used for the majority of our studies since the 4-5-mo Fischer rat had completed most of its growth and was only gaining little weight eating an amount of chow which was little more than the old animals who also were not losing weight. The intestine of these animals showed no evidence of infiltration with inflammatory cells and the proximal small intestine showed similar numbers of bacteria at the two ages (Holt et al, 1985).

Histologic Changes in the Small Intestine of Young and Aging Rats:

Our initial studies showed that the duodenum and jejunum of young and aging Fischer rats did not differ with respect to intestinal dimensions or the number of villus epithelial cells (Holt et al, 1984). However, villus cell number in distal small intestine were greater in aging rats, implying that luminal nutrients reached that area of the gut and induced a greater degree of ileal hyperplasia. This observation suggests that some proximal intestinal dysfunction was present in the senescent rats. When the number of villus cells in proximal intestine of young and aging animals was determined following two or three days of starvation or one day of refeeding after three days fasting, no significant differences were found, implying that the number of functioning villus epithelial cells are well maintained during such nutritional manipulation in the 27-mo-old Fischer rat (Holt and Kotler, 1987).

Surprising, however, was the observation that crypt depth was larger in the proximal intestine of aging rats (Holt et al, 1984). Furthermore, the total number of crypt cells also was greater if one assumes that the crypt has the shape of a cylinder. This increase in crypt cellularity in 27-mo Fischer animals was maintained during three days of starvation and following one day of refeeding (Holt et al, 1988). These observations suggested that increased proliferative activity was ongoing in the older animal. This possibility was supported by findings that the crypt content of thymidine kinase was significantly greater in aging than in young control rats (Holt et al, 1985).

Studies of Intestinal Enzyme Activities in Young and Aging Rats:

In an attempt to evaluate possible changes in epithelial cell function or differentiation, we studied mucosal specific and total enzyme activity of a series of small intestinal epithelial cell enzymes that attain maximal activity in well differentiated villus cells. In these studies, we found either that specific or total activity of proximal intestinal sucrase-isomaltase, lactase and maltase, as well as alkaline phosphatase, was lower in older animals. This was not due to a failure of the small intestinal epithelium of the aging animals to attain enzyme activity as high as younger animals. Instead, using a technique of slicing of cryostat frozen intestinal sections from villus tip to crypt base, we demonstrated that maximum specific activity was similar in the two groups of animals; however, the older rat proximal intestine reached maximum enzyme activity higher up the villus than younger animals (Holt et al, 1985). This "delay in differentiation" appeared to be responsible for lower mucosal enzyme specific activity found in older rats. When the nutrient intake of these animals was altered by 2 or 3 days of starvation, enzyme activity fell in both young and aging animals but the nadir of activity found in the older rats was very much lower than in the young (Holt and Kotler, 1987). Furthermore, after one day of refeeding following 3 days of starvation, there appeared to be an "overshoot" of proximal intestinal enzyme activity of older animals (Holt and Kotler, 1987). These combined results indicate that the 27-mo Fischer rat shows some loss of homeostatic controls which usually maintain the response of enzyme activity during changes in food intake.

Studies of Intestinal Epithelial Cell Proliferation in Young and Aging Rats:

A series of experiments were performed to determine whether crypt cell production might be increased in aging animals. ^{3}H-thymidine labelling of nuclear DNA of crypt cells showed that jejunal crypts of 27-mo Fischer rats contained about 45% more labeled cells than the younger animals, suggesting more proliferative capacity.

To measure crypt proliferation more directly, crypt cell production rate was determined in young and aging small intestine, using the vincristine-induced metaphase accumulation technique (Wright and Appleton, 1980). The rate of accumulation of metaphases per 100 crypt cells, which denotes cells that are cycling, was greater in aging rat small intestine than in the young controls (Holt and Yeh, 1986). Since the crypts of aging animals also were larger, i.e., contained more cells than younger animals, crypt cell production rate (calculated as a product of cells that are cycling and the crypt cell number) was much greater in aging than younger animals (Table 1).

TABLE 1. Effect of Aging upon Small Intestinal Crypt Cell Production Rates

		Control	Starved	Refed
		Crypt cell production rate (cells per hour)		
Duodenum	Y	38.1	22.0	36.3
	A	49.5	44.3	50.8
Jejunum	Y	26.4	ND	26.9
	A	48.8	ND	43.7
Ileum	Y	27.8	17.8	ND
	A	56.7	33.2	ND

Starved rats were starved for 3 days, refed rats were fed for 1 day after the 3rd day of starvation. ND = not determined.
Y = female rats 3-4 mo of age; A = female rats 26-28 mo of age.

The pattern of cell labeling in crypts, 1 hr after administration of ^{3}H-thymidine, reflects the pattern of proliferating cells. Aging rat proximal intestinal crypts demonstrated a broadened zone of crypt proliferation when compared to young intestine, a finding which was particularly apparent following refeeding. A similar pattern of crypt labeling has been found in the normal appearing colon of human patients (Deschner et al, 1963) and relatives of patients (Deschner and Lipkin, 1975) with familial polyposis, in ulcerative colitis (Eastwood and Trier, 1973) and also in the pre-neoplastic stage of chemically induced carcinogenesis of the small and large intestine (Thurnherr et al, 1973; Tutton and Barkla, 1976).

Our preliminary studies of epithelial cell proliferation in the colon also have demonstrated an increase in crypt cell numbers and in the cell production rate in senescent Fischer 344 rats. Difference in proliferation between young and aging rats was particularly evident following three days of starvation when crypt cell production was reduced 4-fold in 3-mo-old rats but by only 20% in 26-28-mo animals. In addition, the pattern of distribution of metaphase accumulation in colonic crypts following vincristine-induced metaphase arrest also suggests the presence of a broadened crypt cell proliferating zone in older animals.

Effects of Diet Restriction on Age-related Changes in the Rat Small Intestine:

Our laboratory has had the opportunity of studying several of our observations related to age-associated changes in gastrointestinal physiology in conventional ad lib fed and 60% dietary restricted (DR) rats who were raised under the auspices of Dr. Arlan Richardson (Chapter 32) For these studies, Fischer 344 male animals were fed a semi-synthetic diet (TekLad Test Diets #5575, Madison WI), ad lib and DR, modeled on the diet extensively studied by Masoro's group in San Antonio (Chapter 3). In Masoro's hands, the median life span of DR rats is extended from about 24 to about 35 mo and the maximum life span from 31 to 42 mo. Dr. Richardson's initial studies were performed in ad lib fed rats sacrificed at 4-5 mo, 12-13 mo, approximately 21 mo and 27-28 mo. DR animals were sacrificed at similar times but, since these rats lived longer, an additional group was studied at 33-34 mo. At the time of sacrifice, Dr. Arlan Richardson's group provided a piece of duodenum, distal to the entrance of the bile duct, and ileum, approximately 6 cm proximal to the ileocecal valve, for histologic study and, for biochemical studies, a 6 cm segment of jejunal mucosa, starting beyond the ligament of Treitz, and of ileal mucosa, 7-13 cm proximal to the ileocecal valve. The intestinal mucosa was scraped and then frozen rapidly. Duodenal villus height in ad lib fed rats varied from 96 to 120 cells and in DR rats from 96 to 118 cells, confirming the absence of a change in villus cellularity with advancing age in either group. In contrast, villus cell numbers in distal ileum increased in ad lib fed animals from 46 to 63 between 4 and 28 mo. These changes did not occur in the ileum of DR rats to 28 mo although ileal height at 33 mo rose by 12% These observations imply that, if proximal intestinal dysfunction occurs during the process of aging and results in ileal hyperplasia, this change is retarded by food restriction.

Duodenal crypt cellularity increased significantly in the 21-22-mo and the 27-28-mo ad lib fed rats. In striking contrast, duodenal cell crypt numbers did not increase in the DR rats from 4 to 28 mo and rose by only 15% at 33 mo over that found at 28 mo in DR animals. If this increase in crypt cell numbers reflects the rate of crypt proliferation, then DR also appeared to retard crypt proliferative changes so prominent in _ad lib_ fed rats.

Finally, jejunal mucosal specific activity changes in the brush border disaccharidases was studied in the two groups of animals. Duodenal lactase activity was consistently higher in the DR animals but fell throughout the age span in both _ad lib_ and DR animal groups. Specific activities of sucrase, maltase and alkaline phosphatase were maintained at significantly higher levels in the DR animals at 27-28 mo than the activities found in _ad lib_ fed animals. By 33 mo, the specific activity of alkaline phosphatase and maltase fell significantly from levels seen at 27-28 mo in the DR group of rats.

CONCLUSIONS

Our studies imply that proliferating tissues, such as the small intestine and colon, are affected by the aging process. Age related change in enzyme synthesis and/or enzyme structure results in lower activity of crucial intestinal enzymes such as brush border disaccharidases and alkaline phosphatase as well as a delay in cellular differentiation. Under conditions of nutritional stress, there is evidence for decontrol of enzyme homeostasis.

Age-associated increases in crypt proliferation are seen in the small intestine and colon of the _ad lib_ fed Fischer 344 rat. Under the stress of nutritional manipulation, evidence has been obtained for a derangement of cellular proliferation which has at least some features characteristic of the pre-cancerous bowel.

Lifelong restriction of caloric intake retards small intestinal biochemical and anatomic changes associated with delayed differentiation and increased proliferation as well as the biochemical changes in brush border enzymes associated with delayed differentiation. These age-related changes at the intestinal level must be taken into consideration when formulating a universal hypothesis of the effects of food restriction upon the aging process.

REFERENCES

Corazza GR, Frazzoni M, Gatto MRA, Gasbarrini G (1986). Aging and small-bowel mucosa: A morphometric study. Gerontology 32:60-65.
Deschner EE, Lewis CM, Lipkin M (1963). In vitro study of human rectal epithelial cells. I. Atypical zone of H3-thymidine incorporation in mucosa of multiple polyposis. J Clin Invest 42:1922-1928.
Deschner EE, Lipkin M (1975). Proliferative patterns in colonic mucosa in familial polyposis. Cancer 24:413-418.
Eastwood GL, Trier JS (1973). Epithelial cell renewal in cultured rectal biopsies in ulcerative colitis. Gastroenterology 64:383-390.
Ecknauer R, Vadakel T, Wepler R (1982). Intestinal morphology and cell production rate in aging rats. J Geront 37:151-155.
Fry RJM, Lesher S, Kohn HI (1961). Age effect on cell-transit time in mouse jejunal epithelium. Am J Physiol 201:213-217.
Hohn P, Gabbert H, Wagner R (1978). Differentiation and aging of the rat intestinal mucosa. Mech Ageing Dev 7:217-226.
Holt PR, Pascal RR, Kotler DP (1984). Effect of aging upon small intestinal structure in the Fischer rat. J Geront 39:642-647.
Holt PR, Tierney AR, Kotler DP (1985). Delayed enzyme expression: A defect of aging rat gut. Gastroenterology 89:1026-1034.
Holt PR, Yeh K-Y (1986). Disordered crypt cell production in aging rat small intestine. Gastroenterology 90:1463.
Holt PR, Kotler, DP (1987). Adaptive changes of intestinal enzymes to nutritional intake in the aging rat. Gastroenterology 93:295-300.
Holt PR, Yeh K-Y, Kotler DP (1988). Altered controls of proliferation in small intestine of the senescent rat. Proc Natl Acad Sci USA 85:2771-2775, 1988.
Kendall MJ (1970). The influence of age on the xylose absorption test. Gut 11:495-501.
Lesher S, Fry RJM, Kohn HI (1962). Influence of age on transit time of cells of mouse intestinal epithelium. I. Duodenum. Lab Invest 10:291-300.
Lipkin M (1981). Proliferation and differentiation of gastrointestinal cells in normal and disease states. I Johnson LR (ed): "Physiology of the Gastrointestinal Tract" New York: Raven Press, pp 145-168.

Phillips RA, Gilder H (1940). Metabolism studies in the albino rat: the relation of age, nutrition and hypophysectomy on the absorption of dextrose from the gastrointestinal tract. Endocrinology 27:601–607.

Thrasher J (1967). Age and the cell cycle of the mouse colonic epithelium. Anat Rec 157:621–626.

Thurnerr N, Deschner EE, Stonehill EH, Lipkin M (1973). Induction of adenocarcinomas of the colon in mice by weekly injections of 1,2-dimethylhydrazine. Cancer Res 33:940–945.

Tutton PJM, Barkla DH (1976). Cell proliferation in the descending colon of dimethylhydrazine treated rats and in dimethylhydrazine induced adenocarcinomata. Virch Arch B Cell path 21:147–160.

Webster SGP, Leeming JT (1975). The appearance of the small bowel mucosa in old age. Age and Ageing 4:168–174.

Webster SGP, Leeming JT (1975). Assessment of small bowel function in the elderly using a modified xylose tolerance test. Gut 16:109–113.

Wright NA, Appleton DR (1980). The metaphase arrest technique. Cell Tissue Kinet 13:643–663.

FATTY ACID COMPOSITIONAL CHANGES IN GERMFREE AND CONVENTIONAL YOUNG AND OLD RATS.

G. Bruckner and K. Gannoe-Hale

Department of Clinical Nutrition, University of Kentucky, Lexington, Kentucky

INTRODUCTION

Specific tissue fatty acid composition, with regard to aging, has received little investigative effort. Polyunsaturated fatty acids (PUFA) oxidation products, i.e. lipofucsin, have been implicated in the aging process and may be related to the tissue content of these fatty acids.

Germfree rats (GF) have increased fecal polyunsaturated fatty acids compared to their conventional (CV) counterparts (Demarne, et.al., 1979) and it has been speculated that changes in fatty acid metabolism, brought about by the absence of the gut microflora, may alter the animal's smooth muscle vascular responses to various agonists (Gordon & Bruckner, 1984). Antibiotic decontaminated rats showed marked changes in gastrointestinal tissue fatty acid patterns compared to conventional controls (Bruckner, unpublished data). The antibiotic treated rats compared to the conventional counterparts had significantly less saturated fatty acids (SAT) than PUFA in membrane tissues (cecum, total SAT/total PUFA; .84 CV, .73 GF). Others have shown that the total body fatty acid composition is similar in germfree and conventional rats (Demarne & Sacquet, 1981), but in rabbits fed low PUFA diets GF vs. CV animals had higher liver levels of arachidonic and linoleic acids (Coates, 1979). It remains unclear whether the fatty acid compositional changes occur only in specific tissues or if the changes which we noted in antibiotic treated animals are a consequence of direct drug effect. Specific tissue fatty acid changes with regard to oxidative stress and aging have received little investigative efforts. Therefore, the purpose of this study was to determine the membrane phospholipid fatty acids composition of gastrointestinal samples derived from germfree and conventional rats of various ages.

MATERIALS AND METHODS

1. Animals and Diets - Weanling male Wistar rats were fed L-485 diets and water ad libitum. The fatty acid composition of the diets as fed are depicted in Table 1. The germfree and conventional animals were maintained under similar environmental conditions from young to old age. Groups of 4 to 6 GF and CV animals were sacrificed at 3, 7 and 30 months of age under halothane anesthesia.

TABLE 1.
FATTY ACID COMPOSITION OF THE DIET FED TO GERMFREE AND CONVENTIONAL YOUNG AND OLD RATS

Fatty Acid	Wt%
16:0	12.4
16:1	.3
18:0	3.5
18:1	23.4
18:2	52.9
18:3	6.8

2. Sample Preparation - Intestinal segments (jejumum and cecum) from the 3, 7 and 30 month old GF and CV animals were quickly removed and placed in hexane: isopropanol (2:1) with .02% BHT added (3-4/group). The samples were stored at -70 degrees C for subsequent fatty acid analysis. The tissue was homogenized and the lipids extracted as previously described (Bruckner et.al., 1984). Phospholipids were separated from neutral lipids using Sep Pak silica cartridges (Juaneda & Rocqueling, 1985). The phospholipid and neutral lipid fractions were saponified and methylated using boron trifluoride in methanol and the fatty acid methyl esters identified by their

retention times compared to standards using Silar 10C packed columns (Bruckner, 1984).

RESULTS-DISCUSSION

The neutral lipid fatty acid composition of the jejunal and cecal tissues was not markedly different between GF and CV rats (data not presented). These observations are similar to those by Demarne & Sacquet(1981). However, the phospholipid fatty acids showed differences due to age as well as the microbial status of the animals (Tables 2, 3 and 4).

TABLE 2.
PHOSPHOLIPID FATTY ACID COMPOSITION OF TISSUES FROM 3 MONTH GERMFREE AND CONVENTIONAL RATS

Fatty Acids	Jejunum CV	Jejunum GF	Wt% Means	Cecum CV	Cecum GF
16:0	20.5	16.3[a]		21.1	21.7
18:0	15.5	11.2		12.4	15.2
18:1	14.9	18.8		16.3	18.4
18:2	21.5	34.2[a]		18.8	19.8
18:1	1.4	1.2		1.3	1.1
20:3	1.5	.9		1.6	.8
20:4	17.7	12.9		16.6	19.4[a]
24:1	3.7	1.9[a]		5.0	4.8
22:5	.8	.3		.9	.6
22:6	2.8	1.8		4.0	2.0[a]
Total PUFA	45.7	51.3		43.2	42.7

a = GF significantly different from CV at $p < .05$.

Germfree vs. conventional animals had significantly more 18:2 n6 and total PUFA in the jejunal phospholipids at 3, 7 and 30 months of age. In the cecal tissue the amounts of 18:2 n6 were only elevated in the 30 month GF compared to CV animals (24.5 wt% vs. 18.1 wt% respectively). These changes in cecal 18:2 n6 were inversely related to the amounts of 20:4 n6. It appears that as the level of 18:2 n6 increases in the GF cecum the level of 20:4 n6 decreases with age (3 mos. vs. 30 mos. GF cecal, 20:4 n6, 19.4 vs. 15.5 wt%

respectively). The amounts of 20:4 n6, which serves as one of the eicosanoid precursors, is elevated in 3 and 7 month GF vs. CV rats but not in the 30 month old animals. Germfree vs. CV rats also had significantly less 16:0 in jejunal phospholipids at all ages; cecal levels of 16:0 in the 7 month GF animals were also lower. These changes may reflect a greater availability

TABLE 3.
PHOSPHOLIPID FATTY ACID COMPOSITION OF TISSUES FROM 7 MONTH GERMFREE AND CONVENTIONAL RATS

Fatty Acids	Jejunum CV	Jejunum GF	Cecum CV	Cecum GF
16:0	22.2	18.6[a]	18.8	11.7[a]
18:0	17.7	13.3[a]	12.6	13.4
18:1	16.2	15.7	17.8	16.1
18:2	17.7	25.3[a]	23.9	20.7
18:3	.3	1.7[a]	.9	2.4[a]
20:3	1.1	1.0	1.1	3.1[a]
20:4	19.3	15.5[a]	16.2	20.1[a]
24:1	.4	3.4[a]	3.7	5.5
22:5	.5	.6	.9	.9
22:6	2.1	3.2	1.9	2.4
Total PUFA	41.0	47.3	44.9	49.6

Wt. Means

a = GF significantly different from CV at p< .05.

of PUFA fatty acids in GF rodents. The total monosaturated fatty acids (18:1 and 24:1) decreased with age in the jejunal phospholipids of the CV rats but were not significantly altered in GF animals; cecal levels of monounsaturated fatty acids (MONO) were not different. The impact of these intestinal tissue fatty acid alterations on subsequent eicosanoid formation are, as yet, unknown. However it is conceivable that biosynthesis of specific eicosanoids may be altered by the fatty acid changes observed in GF rats. Additionally the increased tissue PUFA content might contribute to increased oxidative stress. The PUFAs which are of

TABLE 4.
PHOSPHOLIPID FATTY ACID COMPOSITION OF TISSUES FROM 30 MONTH GERMFREE AND CONVENTIONAL RATS

	Jejunum		Cecum	
Fatty Acid Wt %	CV	GF	CV	GF
16:0	21.5	14.4[a]	18.0	17.5
18:0	19.0	10.9[a]	13.9	11.9
18:1	13.8	20.3[a]	15.5	19.5[a]
18:2	17.9	35.6[a]	18.1	24.5[a]
18:3	.4	.6	.1	.3
20:3	1.4	.9	1.2	.6
20:4	21.2	11.8[a]	18.1	15.5
24:1	.1	1.7[a]	4.0	3.2
22:5	.7	.4	.8	.5
22:6	2.8	1.7	2.1	2.0
Total PUFA	44.4	51.0	40.4	43.4

a = GF significantly different from CV at $p < .05$.

particular interest are those derived from linoleic acid. This essential fatty acid, following desaturation and elongation to arachidonic acid can give rise via cycloxygenase and lipoxygenase to a group of extremely bioactive compounds collectively termed eicosanoids. Depending on the species of eicosanoid produced a variety of physiological responses may be altered, e.g. metarteriole dilation and constriction (Higgs, 1982). The amount and type of precursor fatty acid is therefore extremely important in determining which eicosanoids will be formed. For example, the ratio of n3 to n6 fatty acids can markedly influence the amounts of vasoconstrictor or dilator prostaglandins (PG) formed (Bruckner et al, 1984). One might speculate that the gut microflora alters prostaglandin homeostasis as follows: 1) by degrading PUFAs in the gut lumen and thereby decreasing PG substrate availability, 2) by altering the rate of fat and cholesterol absorption, 3) by degrading endogenously produced prostaglandins, e.g. PGI_2, known to be a vasodilator and, 4) by producing branched chain and short chain volatile fatty acids which might alter the energy requirements of the host. Furthermore, the increased levels of PUFA in the GF vs.

CV rodents may increase the animals requirements for dietary antioxidants. It is obvious that much investigative effort needs to be expended to answer these intriguing questions.

REFERENCES

Bruckner G, Lokesh B, German B, Kinsella JE (1984). Biosynthesis of prostanoids, tissue fatty acid composition and thrombotic parameters in rats fed diets enriched with docosahexaenoic (22:6n3) or eicosapentaenoic (20:5n3) acids. Thromb Res 34:479-497.

Coates ME (1979). Nutrition and metabolism in the gnotobiotic state. In Fliedner T, Heit H, Niethammer D, Pflieger H (eds): "Clinical and Experimental Gnotobiotics, Zentralblatt fur Bakteriologie," Stuttgart, New York: Gustav Fischer Verlag, pp. 29-37.

Demarne Y, Sacquet E (1981). Comparative study of fatty acids from body and liver lipids in germ-free and conventional growing rats. Nutr Report Intl 23:1095-1104.

Demarne Y, Sacquet E, Lecourtier MJ, Flanzy J (1979). Comparative study of endogenous fecal fatty acids in germ-free and conventional rats. Amer J Clin Nutr 32:2027-2038.

Gordon HA, Bruckner G (1984). Anomalous lower bowel function and related phenomena in germ-free animals. In Coates ME, Gustafsson BE (eds): "The Germ-free Animal in Biomedical Research," London: Laboratory Animals LTD, pp. 193-213.

Higgs GA (1982). Prostaglandins and the microcirculation. In Herman AG, Vanhoutte PM, Denolin H, Goossens A (eds): "Cardiovascular Pharmacology of the Prostaglandins," New York: Raven Press.

Juaneda P, Rocqueling JP (1985). Rapid and convenient separation of phospholipids and non-phosphorus lipids from rat heart using silica cartridge. Lipids 20:40-41.

NUTRITIONAL BIOCHEMISTRY

FUNCTIONAL AND BIOCHEMICAL PARAMETERS IN AGING LOBUND-WISTAR RATS

Bernard S. Wostmann, David L. Snyder, Margaret H. Johnson and Shi Shun-di.

Lobund Laboratory, Univ. of Notre Dame, Notre Dame, IN 46556, USA

Dietary restriction has fascinated nutritionists since the early studies of McCay (McCay et al., 1935), and up to this moment we understand little of its basic effects. In 1948 the Dutch government asked the then Netherland Institute of Nutrition at the University of Amsterdam what, in case of World War III, would be the minimum food intake required to keep the Dutch population in acceptable condition. This resulted in studies which indicated that Wistar rats fed 70 % of ad lib. adult intake became lean and mean, but were very healthy. They reproduced well, and outlived their ad lib. fed counterparts by a substantial margin (Van der Rijst et al., 1955). The Amsterdam experience lead to the 30 % reduction in dietary intake used in the present studies.

The emphasis on the germfree (GF) rat in the present investigation is based on the fact that the GF state makes clean "endpoint studies" possible: no microbial infection is going to affect the declining function of the aging animal. It was recognized early that the GF animal is an excellent model for studies in e.g. radiation, lethal trauma, and aging. However, differences in function and metabolism between GF rodents and their conventional (CV) counterparts had become apparent, and the question to what extent these affect results obtained with the GF model cannot be ignored (Wostmann, 1975).

This paper will review the main similarities and differences between male GF and CV Lobund Wistar (L-W) rats maintained on natural ingredient diet L-485 (see D.L Snyder, These Proceedings, chapter 4). This will establish a background against which the results of dietary restriction can then be evaluated.

Subsequently we present data on the effect of dietary restriction on certain metabolic parameters of the aging male L-W rat, in part following suggestions by Masoro (Masoro, 1985). They include: muscle protein turnover, cell size of the liver, and serum lipids; also included are parameters indicating possibly damaging effects of peroxidation: serum malondialdehyde and brain lipofuchsin.

Methods

Animals, housing, diet and maintenance have been described elsewhere in this symposium, as have been methods of sacrifice and necropsy, and the methodology used for the determination of the various hormones (Snyder and Towne, These Proceedings, chapter 14). Serum lipid determinations were part of a study of blood chemistry parameters executed by a Technicon SMAC (Computerized Sequential Multiple Analyzer) at the SouthBend Medical Foundation. Data presented here on serum hormones and blood chemistry came from all available L-W rats examined during the Lobund Aging Project. All data are given $\pm$S.E.

Xylose absorption: After the administration of 450 mg xylose in water by stomach tube, urine was collected for periods up to 24 hours. Xylose in urine was determined with the orcinol color reaction (Ashwell,1957).

3-Methylhistidine (MEH): Urine collection took place in plastic metabolism cages (Nalgene Products). All collections from GF rats were done in plastic isolators under gnotobiotic conditions. After an adjustment period of 3 days, urine was collected for 3 days. The samples were kept at - 20 C until analysis. After a 22 hour hydrolysis with 5 N HCl, the samples were analyzed by a modification of the technique described by Rahda and Bessman (Rahda and Bessman, 1982). This calls for removal of HCl by evaporation, uptake of the sample in 0.1 M phosphate buffer adjusted to pH = 5, and removal of insoluble material by centrifugation. Subsequently the material is brought on a Dowex-50W column. The column is then washed with citrate/phosphate buffer (pH = 5) and eluted with 0.1 M $NaHCO_3$. After evaporation to dryness the samples are dissolved in 0.045 M citric acid and adjusted to pH = 5. After addition of the o-phtalaldehyde/ninhydrin reagent described in the original publication, the color is developed for 10 min at 45 C. The solutions are cooled and read in a spectrophotometer at 490, 450 and 410 nm. The reading at 490 nm (OD490) represents 3-

methylhistidine. Readings at 450 (OD450) and 410 (OD410) nm represent possible interfering materials. They are each deducted from OD490 after recalculating their contribution at 490 nm. Results were calculated based on a standard 3-methyl-1-histidine preparation (Sigma M-3879). They were expressed as mg/d and mg/kg/d 3-methylhistidine excreted in the urine.

DNA in liver and muscle: DNA was determined in the saline homogenates of fresh liver samples following the technique of Prasad et al. After treatment with RNAase, the adduct with ethylene bromide was measured spectrophotofluoremetrically at 320/587 nm (Prasad et al.,1972).

Malondialdehyde (MDA): MDA and related materials were determined spectrophotofluorometrically using in principle the methodology described by Yagi (Yagi, 1976). A fresh serum aliquot is reacted with 0.6 % thiobarbituric acid. After heating for 20 minutes at 100 C, the mixture is cooled, the chromophore extracted with butanol, and fluorescence read 515/553 nm against a standard of 1,1,3,3-tetraethoxypropane.

Lipofuchsin in brain: Brain was homogenized in 20 vol. chloroform-methanol 2:1. After washing the solvent was removed and the residue, after purification, taken up in chloroform-methanol 1:9, brought on a Pharmacia QVF Sephadex K15/30 column, and eluted with the same solvent. The eluted fraction were read with a spectrophotofluorometer at 365/435 nm (Csallany and Ayaz, 1976).

Results

Effects of the germfree state.

Table 1 brings together some of the pertinent differences between GF and CV rats observed in the present and in earlier studies. They suggest that the early recognized low cardiac output and resting oxygen consumption of the young and young adult GF rat (Wostmann, 1975), the low food consumption of the young GF animal, and its always lower than CV heart and liver weights may be related to a small but significant increase in lifespan. The differences in intestinal bile acid composition and concentration, and the obvious difference in absorptive capacity (Table 4; confirming earlier reports (Heneghan, 1963)), all suggest differences that may be of potential importance when specific aspects of function and metabolism are studied.

Table 1. PHYSIOLOGICAL AND METABOLIC PARAMETERS OF THE YOUNG ADULT GERMFREE MALE LOBUND WISTAR RAT

Body weight: same   Body temperature: same
Food intake: less

Percent of conventional value

Liver weight	79	s*	Serum: T_4	116	s
Heart weight	83	s	T_3	117	s
Cardiac output	68	s	Epinephrine	83	
Oxygen use	80#	s	Norepinephrine	102	

Urine: 3-Methylhistidine 96

* indicates a significant difference.
# Sprague-Dawley 74%, Fisher 85 %.

However, notwithstanding these differences, it appears that homeostatic control, to the extent that it is expressed by the levels of various hormones, is mostly comparable between the 2 groups. The data in Table 2 indicate that, of the serum hormone

TABLE 2. HORMONES IN SERUM OF YOUNG ADULT GERMFREE AND CONVENTIONAL RATS

Males 4 -12 months old; diet L-485

	GERMFREE		CONVENTIONAL	GF/CVx100
Body weight, g	376±4		387±5	97
Hb., OD at 413 x 1000	937±22	s	891±14	105
T4, µg/dl	4.92±0.14	s	4.54±0.09	108
T3, ng/dl	111±4	s	89±3	125
Insulin, µIU/ml	35.1±2.5		35.5±1.4	99
Testost., ng/ml	3.17±0.22		3.24±0.32	98
Prolac., ng/ml	23.2±1.3	s	30.5±2.2	76
Gastrin, pg/ml	64±6	s	168±13	38
Epineph., µg/l*	5.76±0.66		6.93±0.74	83
Norepi., µg/l*	7.13±0.95		6.98±0.79	102

*(Sewell and Wostmann, 1976)

levels determined this far, only gastrin is substantially lower in the GF rat. The slighly but consistently higher levels of serum T3 and T4 in the GF animal may result, at least in part, from the somewhat higher hemoconcentration observed in the GF rat, although this will not fully explain the higher T3 values. Otherwise no appreciable differences were found between GF and CV rats.

Effects of lifelong dietary restriction.

Table 3 compares the effects of a 30% dietary restriction, observed in the study at the Netherland Institute of Nutrition, with the present data. Although the diets were completely different (the Dutch experiment reflected the composition of the diet of the Dutch population at the time), in both cases we see an increase in life span of approximately 1/2 year.

Table 3. EFFECT OF A 30 % RESTRICTION OF DIETARY INTAKE ON THE LIFE SPAN OF THE MALE WISTAR RAT

Median life span in months

	DIET	FULLFED	RESTRICTED
Amsterdam 1955	RAN	21.0	25.6
	CV	27.9	
Lobund conventional	L-485	31.2	39.0
germfree		33.6	37.8

RAN: (RAtion Netherlands) was a diet composed of ingredients consumed by the average Dutch population in the years after WW II.
CV: rat diet comparable to L-462 (Wostmann, 1975).

In the present study a number of functional and metabolic parameters were determined to gauge the effects of restriction on metabolism. It should be born in mind that our choices were influenced by the need to use non-invasive techniques as much as possible.

The data in Table 4 show that the GF rat has a much higher capability to absorb xylose than its CV counterpart, although it is recognized that some of the effect illustrated here may be caused by the loss of xylose due to bacterial degradation in the upper small intestine of the CV animals. This absorption is thought to be in part passive, in part carrier-facilitated.

Table 4. URINARY EXCRETION OF XYLOSE DURING 12 HOURS AFTER ADMINISTRATION OF 450 MG BY GAVAGE TO MALE RATS

Percent of dose administered

	GERMFREE		CONVENTIONAL	
	F	R*	F	R*
3 - 12 mo.	33.1±5.6	43.8±5.1	6.9±0.8	11.7±1.3
20+ mo.	50.5±6.9	ND	10.2±1.2	12.5±0.7

* R > F (ANOVA, P = 0.01).

Since carrier-facilitated absorption would depend on genetic make-up and substrate availability, which should be comparable in all groups, it would seem that most of the differences indicated in Table 4 result from differences in the passive absorption capacity of the gut. The data not only confirm the higher absorptive cacacity of the GF gut reported earlier (Heneghan, 1963), but also indicate that dietary restriction increases that capacity, while aging, if anything, does not affect it in a negative way.

Myofibrillar protein turnover, representing mostly the turnover of skeletal muscle and intestinal wall smooth muscle protein, was estimated from the urinary excretion of 3-methylhistidine. It was found to be surprisingly stable over the lifetime of the rat, except for a period of relatively high turnover in very young rats (Table 5). No difference was found between GF and CV rats, but the restricted animals consistently excreted 14 % less per kg/d than fullfed rats. Taking into account the lower amount of body fat and relatively somewhat higher muscle protein content of the restricted animals, the actual difference could be as high as 18 %.

Table 5. EFFECT OF RESTRICTED FEEDING ON THE EXCRETION
OF 3-METHYLHISTIDINE IN URINE

Age 5-30 months; mg/kg/day

Germfree	1.17±0.03	
Conventional	1.13±0.03	P=0.64
Fullfed	1.22±0.03	
Restricted	1.05±0.03 (86 %)	P< 0.01

Liver and muscle DNA and protein content were determined to obtain an impression of the effect of restriction on cellsize. No effect of either microbial status or restriction was found in the muscle values thus obtained. The low DNA content of the liver of the restricted rats was unexpected (Table 6). They contained approximately 3/4 as much DNA/g liver as their fullfed counterparts, with no difference between GF-R and CV-R rats.

Table 6. DNA IN LIVER OF ADULT MALE WISTAR RATS, AGE 5 TO 24 MONTHS[@]

	GF-F	GF-R	CV-F	CV-R
DNA, mg/g	3.3±0.1	2.4±0.1*	3.2±0.1	2.5±0.1*
Protein/DNA, mg/mg	59.2±1.2	73.6±0.9*	60.6±2.2	72.2±1.9*
DNA/100 g body wt, mg	6.6±0.1[#]	6.1±0.1*[#]	8.3±0.4	7.7±0.3

* Significantly different from fullfed.
[#] Significantly different from conventional.

DNA: GF-F = CV-F   Protein/DNA: GF-F = CV-F
     F > R   F < R
  GF-R = CV-R   GF-R = CV-R

[@] Twelve to 30 observations per value indicated. There was a slight increase (< 10 %) in DNA values over the 5 to 24 months period.

Restricted rats averaged 404 mg liver/mg DNA against 311 mg

liver/mg DNA for the fullfed rats, implying a difference in the main cellular component of the liver, the hepatocyte. Using DNA content as an indicator of the number of cells, GF-F rats averaged 80 % of the liver cells per unit of body weight found in the CV-F rats, while GF-R rats averaged another 9 % less, suggesting that functionally the GF-R rat liver is particularly efficient.

A similar conclusion can be drawn from the fact that, while in fullfed rats both serum triglycerides and cholesterol increased substantially with age, the rise in cholesterol was less in restricted rats and no increase in triglyceride concentration was seen (Table 7).

Table 7. SERUM LIPIDS OF MALE LOBUND WISTAR RATS

	GERMFREE		CONVENTIONAL	
	F	R	F	R
Cholesterol, mg/dl				
4 - 12 mo.	86+2	78+2	93+1	91+4
24 - 30 mo.	115+4	86+5	142+7	114+2
Triglycerides, mg/dl				
4 - 12 mo.	91+4	73+5	120+5	94+5
24 - 30 mo.	115+12	77+11	167+17	94+5

Obviously the restricted animals, particularly the GF-R rat, funnel less of the available dietary components into the lipid compartment, utilizing more to achieve and maintain body weight. It is doubtful, however, if these low serum lipid values would benefit the rat, since none of the observed pathology could be connected to the increase in serum lipids with age.

While the 30 % restriction in dietary intake obviously affects metabolic parameters, its effect on peroxidation of membrane phospholipids and its sequelae were less clear. Graph 1 illustrates the occurrence of malondialdehyde (MDA) and related materials. Only after approximately 18 months do the values in the serum increase. The increase shows no difference between GF and CV animals. Although more pronounced in the fullfed than in the restricted rats, the difference is less than dramatic. We also determined the occurrence of lipofuchsin in the brain, since a number of reports have suggested a substantial increase with age of this potentially harmful material, presumed to be a

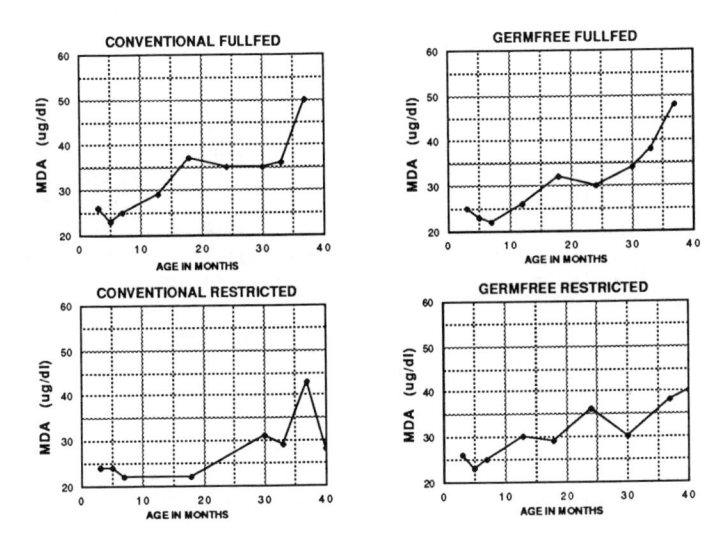

Graph. 1 Malondialdehyde in serum

condensation product of MDA and related materials with lipid compounds carrying free amino groups. We were not able to detect any effect of age, microbial status or dietary intake on the chloroform-methanol extractable fraction which, after further fractionation according to the earlier mentioned technique (see Methods) we defined as "lipofuchsin" (Table 8).

Table 8. LIPOFUCHSIN IN BRAINS OF YOUNG (5 TO 7 MONTHS) AND OLD (30 TO 40 MONTHS) MALE WISTAR RATS MAINTAINED ON DIET L-485
Arbitrary units

	GERMFREE		CONVENTIONAL		ALL
	F	R	F	R	
Young					82+9 (13)
Old	87+9 (10)	82+10 (8)	72+10 (5)	76+12 (7)	80+6 (30)

ANOVA: GF vs CV P= 0.58; F vs R P= 0.66

Discussion.

Our results indicate that under the present experimental conditions the CV rats live almost as long as the GF rats. Kidney disease in the CV rats has been virtually eliminated, presumably because no kidney infection was seen (Snyder and Wostmann, These Proceedings, Chapter x), and the diet contained soy protein instead of animal protein (Iwasaki et al., 1988). It would appear that, if other laboratories would be able to duplicate our experimental conditions, CV rats would equal GF rats as models in aging studies.

Restriction of both our GF and our CV L-W rats still had the usual effect on morbidity and mortality. The question whether restriction decreases or increases metabolic rate remains unsolved (see Jukes vs Masoro, Letter to the Editor, 1987), but our data indicating a decrease in myofibrillar protein turnover would seem to agree with the former supposition. These data also appear in accord with earlier observations (Waterlow, 1986).

References.

Ashwell G (1957) Colorimetric analysis of sugars. In Colowick SP, Kaplan NO (eds): "Methods in Enzymology, Vol 3" New York: Academic Press, pp 73-105.

Connor Johnson B versus Masoro EJ (1986). Letter to the Editor. J Nutr 116:323-325.

Csallany AS, Ayaz KL (1976). Quantitative determination of organic solvent-soluble lipofuchsin pigments in tissues. Lipids 11:412-417.

Heneghan JB (1963). Influence of microbial flora on xylose absorption in rats and mice. Am J Physiol 205:417-420.

Iwasaki K, Gleiser CA, Masoro EJ, McMahan CA, Seo E, Yu BP (1988). The influence of dietary protein source on longevity and age-related disease processes of Fisher rats. J Gerontol 43:B5-12.

Masoro EJ (1985). Nutrition and aging-a current assesment. J Nutr 115:842-848.

McCay CM, Crowell MF, Maynard LA (1935). The effect of retarded growth upon the length of life span and upon the ultimate body size. J Nutr 10:63-79.

Prasad AS, DuMouchelle E, Koniuch D, Oberleas D (1972). A simple fluorometric method for the determination of RNA and DNA in tissues. J Lab Clin Med 80:598-602.

Rahda E, Bessman SP (1982). A rapid colorimetric method for 3-methylhistidine in urine. Analyt Biochem 121:170-174.

Sewell DL, Bruckner-Kardoss E, Lorenz LA, Wostmann, BS (1976). Glucose tolerance, insulin and catecholamine levels in germfree rats. Proc Soc Exp Biol Med 152:16-19.

Van der Rijst MP, Jansen BC, Beeker TW, Wostmann BS (1955). Experiments to determine the nutritive value of the average diet consumed in the Netherlands, when fed to white rats ad libidum and under conditions of restricted consumption (70 %). Voeding 16:708-725.

Waterlow JC (1986). Metabolic adaptation to low intakes of energy and protein. Ann Rev Nutr 6:495-526.

Wostmann BS (1975). Nutrition and metabolism of the germfree mammal. World Rev Nutr Diet 22:40-92.

Yagi K (1976). A simple fuorometric assay for lipoperoxide in blood plasma. Biochem Med 15:212-216.

BLOOD GLUTATHIONE: A BIOCHEMICAL INDEX OF LIFE SPAN ENHANCEMENT IN THE DIET RESTRICTED LOBUND-WISTAR RAT

Calvin A. Lang, Wenkai Wu, Theresa Chen, and Betty Jane Mills
Departments of Biochemistry and of Pharmacology and Toxicology, University of Louisville School of Medicine, Louisville, Kentucky 40292.

Our previous findings indicated that a glutathione deficiency of aging occurs in different tissues from two well-characterized model organisms, namely, the C57BL/6J mouse (Hazelton and Lang, 1980) and the yellow fever mosquito, Aedes aegypti (Hazelton and Lang, 1984). Of special interest is that the blood glutathione (GSH) levels in the mouse paralleled and reflected the levels in less accessible tissues, such as liver, kidney, heart, and lung (Abraham, Taylor, and Lang, 1978). In addition, we found that high blood glutathione values are associated with healthy, long-lived human subjects (Richie and Lang, 1986).

The specificity of this phenomenon and its causal mechanism were demonstrated by correction of the GSH deficiency in the mosquito model. This was accomplished by administration of reducing agents, which spare the need for GSH (Richie, Mills, and Lang, 1986), or of precursors of cysteine and GSH, such as magnesium thiazolidine carboxylate (Richie, Mills, and Lang, 1987). The latter compound not only restored the GSH tissue levels to those of young adult ages, but also increased the median life span by 40-50%, thereby verifying the role of GSH in longevity.

The fundamental question, "What is the GSH status of a long-lived rat ?" was the goal of the current work. Other questions were "Does the aging rat also develop a GSH deficiency like mosquito, mouse, and man ?" and "Do germfree conditions affect the GSH status ?" These were investigated with the collaboration of Dr. David Snyder of the University of Notre Dame who provided tissues from diet restricted and fullfed rats from the Lobund aging rat colonies.

The specific objectives were two-fold. First, what are the survival and aging rates of fullfed and diet restricted Lobund-Wistar rats reared under conventional or germ-free conditions? Second, what are the blood GSH levels during their life spans?

The first question was answered by statistical analysis of the survival data for the four experimental groups by our Teton method (Richie and Lang, 1980). In brief, this analysis transforms the data to ln % mortality for the short-lived subpopulation from 0-50% mortality and to ln % survival for the long-lived subpopulation from 50-0% survival. The resultant plot of these parameters versus age is a sharp peak, hence the name, Teton. Analysis of this plot gives 3 parameters, namely, the median survival age, the aging rate of the short-lived, and the aging rate of the long-lived subpopulations.

The median survival ages of <u>fullfed</u> rats were the same regardless of conventional or germfree conditions. Likewise, the median survival ages of <u>restricted</u> rats were the same. Thus germfree conditions had no effect on longevity. Therefore, the data were combined for both fullfed and also for both restricted groups to simplify subsequent analyses.

The results of the Teton analysis are shown in Table 1. In comparison to the fullfed group the median survival age of the restricted rats was 1170 days, which was 23% greater than the fullfed survival age of 950 days ($P< 0.000002$).

These results are seen more clearly in the upper graph of Figure 1. Also the lower Teton graph of the same survival data indicate that the restriction effect is due to a delay or lag of 8 months in the onset of mortality. In addition, the aging <u>rates</u> of both short-lived subpopulations are the same as indicated by the parallel slopes of the ascending lines.

Table 1. Survival and Aging Rates of Fullfed and Diet Restricted Lobund-Wistar Rats

		Median Survival Age (days)		Aging Rate	
Groups	No. of Rats	Median	95% CL	Short Lived Subpop.*	Long Lived Subpop.
Fullfed Combined	79	950	872 - 1028	5.91	7.97
Restricted Combined	96	1170	1115 - 1225	6.18	12.40
P-Value		0.000002		0.26	0.003

*Short-Lived Subpopulation = 1000 x Ln % Mortality (from 0 to 50+ % Mortality)
Long-Lived Subpopulation = -1000 x Ln % Survival (from 50+ to 0 % Survival)

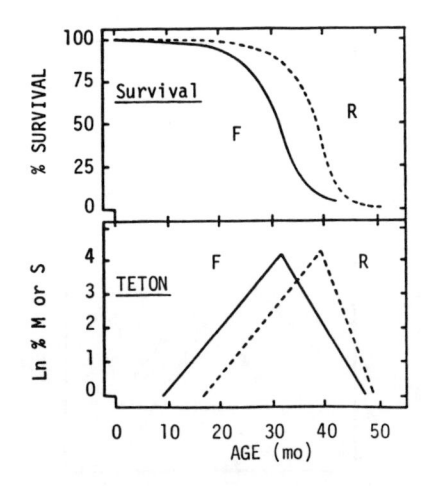

Figure 1. Survival and Teton Curves of Fullfed (F, o———o) and restricted (R, o----o) LOBUND-WISTAR rats.

However, the aging rate of the long-lived, restricted group (dashed line after the peak) was greater indicating that the restricted rats died at a more rapid rate than the fullfed.

The second aspect of this study was the determination of **blood GSH** levels during the life span of the different experimental groups. Whole blood samples were collected at Lobund, quickly frozen in liquid nitrogen, shipped to Louisville in dry ice, and stored at -70°C. They were processed and analyzed for GSH content with our standardized and validated HPLC method with dual electrochemical detection (Richie and Lang, 1987).

The blood GSH profiles from 6-40 mo age are shown in Figure 2. In the upper graph the data demonstrate that approximately 20-100% higher GSH levels occurred at all ages of the restricted compared to the fullfed rats.

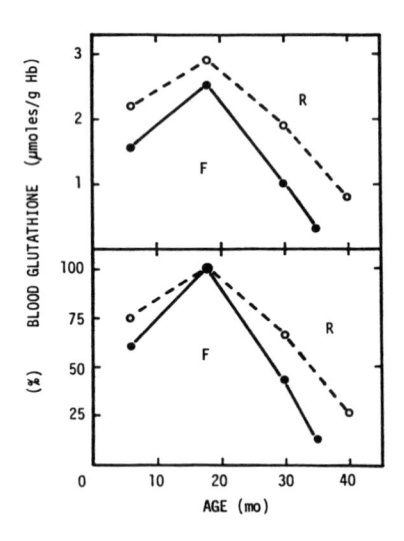

Figure 2. Higher blood GSH levels occur in restricted rats through the life span. F, o———o, fullfed rats, R, o----o, restricted rats.

In the lower graph the data have been expressed as percentage of the maximum at 18 mo. for each group to demonstrate the relative rates of GSH increase and decrease that occurred during the life span. The slopes of these profiles indicated that the rate of GSH increase in the restricted group (dashed line up to 18 mo) is less than in the fullfed. After that maximum, the rate of GSH decrease in the restricted was lower than in the fullfed rats. Thus GSH status in the long-lived restricted rats was maintained longer.

SUMMARY

These experimental results demonstrate that dietary restriction in both conventional and germfree rats results in enhanced longevity compared to fullfed animals. This increase in median survival age was due to a delay of 8 months in the onset of mortality in the restricted rats. Thereafter the aging rates of the short-lived subpopulations were the same for both groups. However, the aging rate of the long-lived restricted subpopulation was greater.

The blood glutathione profiles demonstrated that a GSH deficiency of aging occurred in both the restricted and the long-lived groups. These data confirmed in the rat the findings observed previously in mosquito, mouse and man and verified the generality of the GSH and longevity relationship.

Of special interest is that the blood glutathione levels were consistently higher at all ages of the restricted compared to fullfed rats and decreased more slowly during senescence.

These findings indicate a direct relationship between enhanced GSH status and increased longevity due to dietary restriction. Further this suggests that glutathione may be a molecular mechanism for the diet restriction and longevity phenomenon.

ACKNOWLEDGEMENT

We thank RJR for their generous support of this research.

REFERENCES

Abraham EC, Taylor JF, Lang CA (1978) Influence of mouse age and erythrocyte age on glutathione metabolism. Biochem J 174:819-825.

Hazelton GA, Lang CA (1980) Glutathione contents of tissues in the aging mouse. Biochem J 188:25-30.

Hazelton GA, Lang CA (1984) Glutathione levels during the mosquito life span with emphasis on senescence. Proc Soc Exp Biol Med 176:249-256.

Richie JP Jr, Lang CA (1980) The Teton Method: A simple quantitative analysis of life span profiles. Fed Proc 39:1796.

Richie JP Jr, Lang CA (1986) The maintenance of high glutathione levels in healthy, very old women. Gerontologist 26:80A.

Richie JP Jr, Lang CA (1987) The determination of glutathione, cyst(e)ine, and other thiols and disulfides in biological samples using high-performance liquid chromatography with dual electrochemical detection. Anal Biochem 163:9-15.

Richie JP Jr, Mills BJ, Lang CA (1986) Dietary nordihydroguaiaretic acid increases the life span of the mosquito. Proc Soc Exp Biol Med 183:81-85.

Richie JP Jr, Mills BJ, Lang CA (1987) Correction of a glutathione deficiency in the aging mosquito increases its longevity. Proc Soc Exp Biol Med 184:113-117.

CELLULAR ANTIOXIDANT DEFENSE SYSTEM

Linda H. Chen and Stephen R. Lowry

Department of Nutrition and Food Science, (L.H.C.), and Agricultural Experiment Station (S.R.L.), University of Kentucky, Lexington, Kentucky 40506

INTRODUCTION

Dietary restriction and germfree environment extend life span by unknown mechanism in all species tested. The free radical theory of aging proposed by Harman (1956) hypothesizes that the accumulation of free radical-initiated damage is the major factor in the degenerative changes associated with aging. It is proposed that the aging process is the sum of the deleterious free radical reactions going on continuously throughout the cells and tissues, and these free radical reactions are involved in the pathogenesis of diseases, such as cancer and atherosclerosis (Harman, 1986). Free radical-initiated lipid peroxidation produces tissue damage and accelerates the aging process, thus shortening the lifespan of an organism. It is known that free radical reactions are initiated by enzymatic and non-enzymatic means in living cells, however, their contribution to the process of aging is yet to be determined. The susceptibility of a given tissue to free radical and oxidative damage is a function of overall balance between the degree of oxidative stress and the antioxidant defense capability.

The cell contains various non-enzymatic and enzymatic antioxidant defense systems. Vitamin E is the most important biological free radical scavenger due to its lipid solubility and occurrence in membranes. It reacts as a chain-breaking antioxidant in the inhibition of the free radical peroxidation of polyunsaturated lipids. In addition to the scavenging of free radicals by vitamin E, the cellular enzymatic antioxidant defense system includes superoxide dismutase (SOD), the gluthathione (GSH) peroxidase system and catalase. SOD protects the cell against the deleterious effects of superoxide radicals. Hydroperoxides formed in the cells are decomposed by GSH peroxidase in a reaction in which GSH serves as the substrate. The restoration of GSH involves the reduction of oxidized glutathione in the presence of NADPH with the action of glutathione reductase. Hydrogen peroxide produced at higher concentrations is decomposed by catalase, while hydrogen peroxide at lower concentrations is decomposed by selenium (Se)-dependent GSH peroxidase (Nagel and Ranney, 1973).

This study reports on the status of the cellular antioxidant defense system of germfree and conventional Lobund-Wistar rats, which were full-fed or dietary restricted, at different ages.

MATERIALS AND METHODS

The treatment design is 2 x 2 x 3 (environment x dietary level x age) factorial design conducted in a completely randomized experimental design. Forty-eight Lobund-Wistar rats were divided into 4 groups as follows: conventional-full fed (CF), conventional-restricted (CR), germfree-full fed (GF) and germfree-restricted (GR), and one-third were killed at 7, 18 and 30 months, respectively. Each subgroup consisted of 4 rats.

The heparinized blood was centrifuged to separate plasma and erythrocytes. The liver was removed and 10% liver homogenate in cold 0.1 M phosphate buffer (pH 7.4) was prepared at $0°C$. Cytosol was prepared by centrifuging liver homogenate at 30,000xg for 30 minutes at $0°C$. Plasma samples and liver cytosol samples were frozen at $-80°C$ until analyses could be performed.

Plasma vitamin E levels were determined with a modification of the method of Hashim and Schuttringer (1966) by esterifying the lipid extract of plasma before color production with ferric chloride and bipyridyl reagents. Superoxide dismutase (SOD) activity was assayed by the method of McCord and Fridovich, (1966), and catalase activity was determined by the method of Beers and Sizer (1952). Se-dependent GSH peroxidase activity was determined by the method of Paglio and Valentine (1967) using hydrogen peroxide as the substrate. Protein concentration was determined by the method of Lowry et al. (1951).

All data were subjected to three way analysis of variance. In addition, two way analysis of variance at each age, one way analysis of variance with regard to the effect of age within each group, one way analysis of variance with regard to the effect of environment at each feed level and at each age, and one way analysis of variance with regard to the effect of dietary restriction in each environment and at each age were completed to examine biological relationship from all possible viewpoints.

RESULTS

Table 1 shows the plasma vitamin E levels of the rats. Plasma vitamin E levels increased significantly with age, and decreased significantly with dietary restriction in both conventional and germfree rats at the three ages. Germfree environment significantly increased plasma vitamin E levels at 7 and 30 months in full-fed rats. The three way analysis of variance of the data is shown in Table 2. Age, dietary restriction and germfree environment had

TABLE 1. Plasma Vitamin E Levels[1,2,4]

Environment	Age (Month)	Dietary Treatment Full-fed	Restricted	Level of Significance[3]
Conventional	7	861 ± 49[a]	641 ± 49[a]	P<0.005
	18	1154 ± 49[b]	838 ± 49[b]	P<0.0001
	30	1208 ± 49[b]	991 ± 49[c]	P<0.0001
Level of Significance[3]		P<0.005	P<0.005	
Germfree	7	1004 ± 49[c]	765 ± 49[a]	P<0.005
	18	1187 ± 49[b]	927 ± 49[b]	P<0.001
	30	1407 ± 49[d]	1089 ± 49[c]	P<0.005
Level of Significance[3]		P<0.005	P<0.005	

[1] μg/100 ml plasma.
[2] Mean ± SE of 4 rats.
[3] Based on one way analysis of variance.
[4] Means in each column not sharing a common superscript letter were significantly different (p=0.05).

TABLE 2. Three-Way Analysis of Variance of All Data[1] And Level of Significance

Source	Vitamin E	SOD	Catalase	Se-Dependent GSH Peroxidase
Age	P<0.0001	n.s.[2]	n.s.	n.s.
Restriction	P<0.0001	P<0.0005	n.s.	P<0.0001
Germfree	P<0.0005	P<0.0001	n.s.	P<0.05
Age x Restriction	n.s.	P<0.0001	P<0.0001	n.s.
Age x Germfree	n.s.	n.s.	P<0.005	n.s.
Restriction x Germfree	n.s	n.s.	n.s.	n.s.
Age x Restriction x Germfree	n.s.	P<0.0001	P<0.05	n.s.

[1] Data of Tables 1,3-5.
[2] Not significant.

significant effects on plasma vitamin E levels. There was no significant two way or three way interaction among the three factors.

Table 3 shows liver SOD activity. SOD activity decreased significantly with age within the CF group, however, it increased significantly with age within the CR group. Dietary restriction significantly decreased SOD activity at 7 and 18 months but significantly increased the enzyme activity at 30 months in the conventional rats. Germfree environment significantly decreased SOD activity at each age level in the full-fed and restricted rats. The three way analysis of variance in Table 2 shows that there were significant interactions among the three factors (age, restriction and germfree environment), and also between age and restriction.

In Table 4 liver catalase activity is shown. The enzyme activity decreased significantly with age within the CF group. On the other hand, it increased significantly with age within the GR group. Dietary restriction significantly decreased the enzyme activity at 7 months, but significantly increased the enzyme activity at 18 and 30 months in the conventional rats. In germfree rats, dietary restriction significantly decreased the enzyme

TABLE 3. Liver Superoxide Dismutase Activity[1,2,5]

Environment	Age (Month)	Dietary Treatment Full-fed	Restricted	Level of Significance[3]
Conventional	7	57.7 ± 1.2^a	43.6 ± 1.2^a	P<0.0001
	18	55.2 ± 1.2^a	47.2 ± 1.2^a	P<0.0001
	30	45.1 ± 1.2^b	55.9 ± 1.2^b	P<0.0001
Level of Significance[3]		P<0.005	P<0.005	
Germfree	7	38.5 ± 1.2^c	37.8 ± 1.2^c	n.s.[4]
	18	—	39.3 ± 1.2^c	—
	30	38.9 ± 1.2^c	37.4 ± 1.2^c	n.s.
Level of Significance[3]		n.s.	n.s.	

[1] Units/min./mg protein.
[2] Mean ± SE of 4 rats.
[3] Based on one way analysis of variance.
[4] Not significant.
[5] Means in each column not sharing a common superscript letter were significantly different (P=0.05).

activity at 7 months. Germfree environment had significant effects in the full-fed rats. The three way analysis of variance in Table 2 indicates that there were significant interactions among the three factors, and between age and restriction, and also between age and germfree treatment.

Se-dependent GSH peroxidase activity in the liver is shown in Table 5. The enzyme activity was not significantly affected by age, but was significantly increased by dietary restriction at each age in both conventional and germfree rats. Germfree environment significantly increased the enzyme activity in the full-fed rats. The three way analysis of variance in Table 2 shows that dietary restriction and germfree environment had significant effects on this enzyme activity.

DISCUSSION

In rats, contents of vitamin E in most tissues have been found to increase with age (Weglicki et al., 1969). Lower serum tocopherol levels were found in 24 month-old mice than in 3 month old mice when the dietary vitamin E level was lower than adequate (30 ppm) (Meydani et al.,

TABLE 4. Liver Catalase Activity[1,2,5]

Environment	Age (Month)	Dietary Treatment Full-fed	Restricted	Level of Significance[3]
Conventional	7	1.67 ± 0.07[a]	1.23 ± 0.07[a]	P<0.0001
	18	1.24 ± 0.07[b]	1.41 ± 0.07[b]	P<0.0001
	30	1.17 ± 0.07[b]	1.48 ± 0.07[b]	P<0.0001
Level of Significance[3]		P<0.01	n.s.[4]	
Germfree	7	1.38 ± 0.07[c]	1.04 ± 0.07[a]	P<0.005
	18	—	1.47 ± 0.07[b]	—
	30	1.41 ± 0.07[c]	1.42 ± 0.07[b]	n.s.
Level of Significance[3]		n.s.	P<0.001	

[1-5]Same as Table 3.

TABLE 5. Liver Se-Dependent Glutathione Peroxidase Activity[1,2,5]

Environment	Age (Month)	Dietary Treatment Full-fed	Restricted	Level of Significance[3]
Conventional	7	780 ± 85[a]	1135 ± 85[a]	P<0.01
	18	834 ± 85[a]	1281 ± 85[a]	P<0.005
	30	785 ± 85[a]	1267 ± 85[a]	P<0.0005
Level of Significance[3]		n.s.[4]	n.s.	
Germfree	7	1050 ± 85[b]	1336 ± 85[a]	P<0.05
	18	—	1218 ± 85[a]	—
	30	1030 ± 85[b]	1309 ± 85[a]	P<0.05
Level of Significance[3]		n.s.	n.s.	

[1]nmole/min./mg protein.
[2-5]Same as Table 3.

1986). Studies in humans have shown an age-related increase in serum tocopherol levels through age sixty (Chen et al., 1977; Wei Wo and Draper, 1975) followed by a decline thereafter (Barnes and Chen, 1981; Wei Wo and Draper, 1975). These alterations are probably associated with similar changes in plasma lipid levels (Horwitt et al., 1972). Our results showed an increase in plasma vitamin E levels with increasing age. Because vitamin E is a fat-soluble vitamin, it is stored in the adipose tissue; with the high dietary vitamin E level in this study (243.2 IU/kg diet) the amount of storage is expected to increase with age. In this study plasma vitamin E levels were decreased by dietary restriction. This result could be explained by the decreased vitamin E intake and the possible decreased adiposity with dietary restriction. That plasma vitamin E levels were significantly higher in germfree than in conventional rats may be due to a lower metabolic rate in germfree animals (Wostmann, 1975). It also suggests less free radical reactions in rats raised in the germfree environment which might have led to lower requirement for vitamin E.

Consistent with our results in the CF group, the activity of SOD has been found to decline with age in some studies with rats and mice (Danh et al., 1983; Reiss and Gershon, 1976). However, in other studies with rats (Kellog and Fridovich, 1976) and mice (Koizumi et al., 1987) no significant changes in the enzyme activity with increasing age have been observed.

In a study with mice, dietary restriction has been reported to have no effect on the SOD activity (Koizumi et al., 1987). Our results showed that dietary restriction of about 30% produced a decrease in the enzyme activity at 7 and 18 months, and an increase in the enzyme activity at 30 months of age in the conventional rats. In this study germfree environment significantly lowered the SOD activity suggesting the possibility of less superoxide radical production due to lower metabolic rate and/or lower rate of protein biosynthesis in germfree rats (Muramatsu et al., 1981).

Our results in the CF group were consistent with a study using rats which demonstrated that catalase activity decreases with age (Ross, 1969). A study on the regulation of catalase activity in mice of different ages has suggested that either an increase in the rate of degradation or a decrease in the rate of synthesis of catalase occurs at old age (Baird and Samis, 1971). In a study with mice (Koizumi et al., 1987) and a study with rats (Ross, 1969), dietary restriction has been reported to increase catalase activity. Our results at ages 18 and 30 months in conventional rats were consistent with these studies.

In contrast to our results, a study with mice has shown that GSH peroxidase decreases at old ages (Hazelton and Lang, 1985). A dietary restriction study with mice (Koizumi et al., 1987) has shown that restriction produces no change in the enzyme activity. Dietary restriction in this study resulted in a significantly increased Se-dependent GSH peroxidase activity which suggests augmentation of the enzyme activity by dietary restriction. It is interesting to note that the pattern of Se-dependent GSH peroxidase of the four groups of animals appears to parallel the pattern of the life span of the animals: GR>CR>GF>CF.

Some reports in the literature conflict with our results. This discrepancy may be due to the differences in the species, strain, gender and ages of animals, and the diets used in the study. For example, the control diet of female F_1 hybrid strain mice used by Koizumi et al. (1987) was restricted by about 20% compared to the usual ad libitum-fed control animals, and enzyme activities determined at 12 and 24 months. In the study of antioxidant defense systems, the levels of dietary components, especially of vitamin E and Se have an important influence. In this study, the diet contained a high level of vitamin E. In addition, there may have been other conditions which differed from those of other studies.

The findings that dietary restriction resulted in a higher SOD and catalase activity in the old conventional rats suggest that this life-prolonging treatment provided greater protection against superoxide radical and high concentrations of hydrogen peroxide at old age. In addition, the findings that dietary restriction produced a significant increase in Se-dependent GSH peroxidase activity in all rats also suggest that this treatment resulted in a higher degree of protection against low concentrations of hydrogen peroxide. The mechanism whereby dietary restriction prolongs the life span is not clear. Masoro (1985) has reviewed the hypotheses proposed by various researchers and summarized that food restriction may prolong life span possibly by affecting the secretion of pituitary hormones, by delaying the age-related decline in cellular protein turnover, or by inhibiting free radical-mediated damage. Walford et al. (1987) have suggested that long-term energy restriction may extend the life span by affecting the generation or persistence of free radicals, and that energy restriction leads to a selective up-regulation of certain antioxidant defense systems by mechanisms other than feed back response to DNA damage or to free radical production.

Our study on enzyme activities with dietary restriction appears to be consistent with a role of free radicals in the process of aging. The life prolonging effect of germfree treatment, however, may have different mechanisms. The lower metabolic rate and less infection in the germfree animals may have contributed to the extension of the life span of these animals.

ACKNOWLEDGMENT

The technical assistance of Mei-Qi Zhu is sincerely acknowledged.

REFERENCES

Baird MB, Samis HV (1971). Regulation of catalase activity in mice of different ages. Gerontologia 17:105-115.
Barnes KJ and Chen LH (1981). Vitamin E status of the elderly in Central Kentucky. J Nutr Elderly 1:41-49.
Beers RF, Sizer IW (1952). A spectrophotometric method for measuring the breakdown of hydrogen peroxide by catalase. J Biol Chem 195:133-140.
Chen LH, Hsu SJ, Huang PC, Chen JS (1977). Vitamin E status of Chinese population in Taiwan. Am J Clin Nutr 30:728-735.
Danh HC, Benedett MS, Dostest P (1983). Differential changes in superoxide dismutase activity in brain and liver of old rats and mice. J Neurochem 40:1003-1007.
Harman D (1956). Aging: a theory based on free radical and radiation chemistry. J Gerontol 11:298-300.
Harman D (1986). Free radical theory of aging: role of free radicals in the origination and evolution of life, aging, and disease processes. In Johnson JE, Walford R, Harman D, Miguel J (eds): "Free Radicals, Aging, and Degenerative Diseases," New York: Alan R. Liss, pp 3-49.
Hashim SA, Schuttringer GR (1966). Rapid determination of tocopherol in macro- and micro- quantities of plasma. Am J Clin Nutr 19:137-145.
Hazelton GA, Lang CA (1985). Glutathione peroxidase and reductase activities in aging mouse. Mech Ageing Dev 29:71-81.

Horwitt MK, Harvey CC, Dahn CJ, Searey MT (1972). Relationship between tocopherol and serum lipid levels for determination of nutritional adequacy. Ann N Y Acad Science 203:223-236.

Kellog EW, Fridovich I (1976). Superoxide dismutase in the rat and mouse as a function of age and longevity. J Gerontol 21:405-408.

Koizumi A, Weindruch R, Walford RL (1987). Influences of dietary restriction and age on liver enzyme activities and lipid peroxidation in mice. J. Nutr 117:361-367.

Lowry OH, Rosebrough MJ, Farr AL, Randall RJ (1951). Protein measurement with the Folin phenol reagent. J Biol Chem 193:265-275.

Masoro EJ (1985) Nutrition and aging - a current assessment. J Nutr 115:842-848.

McCord JM, Fridovich I (1966). Superoxide dismutase. J Biol Chem 244:6049-6055.

Meydani SN, Meydani M, Verdon CP, Blumberg JB, Hayes KC (1986). Vitamin E supplementation suppresses prostaglandin E1/2 synthesis and enhances the immune response of aged mice. Mech Ageing Dev 34:191-201.

Muramatsu T, Coates ME, Hewitt D, Salter DN, Garlick PJ (1981). Protein synthesis in liver and jejunal mucosa of germfree chicks. In Suzaki S, Osaei A, Hashimoto K (eds): "Recent Advances in Germfree Research," Tokai Univ Press pp 303-305.

Nagel R, Ranney H (1973). Drug-induced oxidative denaturation of hemoglobin. Seminars in Hematol 10: 269-278.

Paglio DE, Valentine WN (1967). Studies on the quantitative and qualitative characterization of erythrocyte glutathione peroxidase. J Lab Clin Med 70:158-169.

Reiss U, Gershon D (1976). Comparison of cytoplasmic superoxide dismutase in liver, heart, and brain of aging rats and mice. Biochem Biophys Res Commun 73:255-262.

Ross MH (1969). Aging, nutrition and hepatic enzyme activity pattern in the rat. J Nutr 97:563-602.

Walford RL, Harris SB, Weindruch R (1987). Dietary restriction and aging: historical phases, mechanisms and current directions. J Nutr 117:1650-1654.

Weglicki WB, Luna F, Nair PP (1969). Sex and tissue specific differences in concentrations of α-tocopherol in mature and senescent rats. Nature 221:185-186.

Wei Wo CK, Draper HH (1975). Vitamin E status of Alaskan Eskimos. Am J Clin Nutr 28:808-813.

Wostmann BS (1975). Nutrition and metabolism of the germfree mammal. World Rev Nutr Diet 22:40-92.

CELLULAR BIOCHEMISTRY

AN OVERVIEW OF AGE-RELATED CHANGES IN PROTEINS

Morton Rothstein

Department of Biological Sciences

State University of New York at Buffalo
Amherst, New York 14260

The biochemistry of aging covers a wide variety of topics ranging from changes in hormone response to oxidative damage to DNA. Some of the biochemical studies were inspired by various theories of aging. Examples are the Free Radical Theory, the idea that aging is related to the metabolic rate and the Error Catastrophe Hypothesis. Other studies were initiated because they involve basic metabolic processes. Examples are studies of mitochondria, changes in enzyme levels and studies of membrane function. Still other studies were developed because of fortuitous findings that could be deemed important to the aging process. Examples are studies of late passage cells and increases in protein synthesis in the livers of very old rats. Though biochemists have discovered and grouped together many individual facts, they have, on the whole, failed to penetrate deeply into any metabolic process that can be linked directly to aging.

Aging research on proteins received a sharp stimulus from the Error Catastrophe Hypothesis proposed by Orgel (1963; 1970). Orgel proposed that if an error occurred in the protein-synthesizing machinery, it would bring about errors in the proteins being synthesized. If some of those error-containing proteins subsequently became part of the protein-synthesizing machinery, they would cause even more errors in the next generation of proteins. These proteins, in turn, would bring about even more errors. Thus, there would be an __amplification__ of errors until a catastrophe occurred.

The Error Catastrophe Hypothesis stimulated a great deal of discussion and research. It did not take long to dispense with the idea of a "catastrophe", but the concept of "errors" hung on until quite recently, not only in proteins, but in nucleic acids as well. The idea still occasionally surfaces in the introductory sections of papers on aging. However, it now seems clear that errors in molecules, as differentiated from metabolic changes do not cause aging and aging does not cause errors in molecules.

In 1970, it was reported that in the aged, free-living nematode, *Turbatrix aceti*, the enzyme, isocitrate lyase possessed altered properties (Gershon and Gershon, 1970). The "old-type" enzyme showed an altered sensitivity to heat, a lower specific activity and a reduced response to antiserum produced against young nematode homogenate. Though these experiments were carried out with crude preparations, the data were consistent and convincing. Several other crude enzymes were subsequently reported to be altered in old animals and in late passage cells in culture (for review, see Rothstein, 1982).

As can be imagined, these early reports stimulated a good deal of research and particularly, discussions, pro and con, about the Error Hypothesis. In short, there developed a consciousness among researchers that errors could play a role in the aging process.

Eventually, more detailed studies were undertaken utilizing pure enzymes. As with crude enzyme preparations, the pure altered enzymes showed changes in heat-sensitivity, specific activity and in immunological response. Antiserum produced to young-type enzyme responds more efficiently to young than to old enzyme. That is, it requires less antiserum to inactivate "young" enzyme. The reverse is also true if the antiserum is prepared to "old" enzyme.

Enzymes that have been obtained in pure form and without contradiction, have been established as having altered "old" forms are as follows: In nematodes, isocitrate lyase, phosphoglycerate kinase, enolase and aldolase; in rats, phosphoglycerate kinase (muscle, liver, brain), maltase (kidney), glyceraldehyde-3-phosphate dehydrogenase (muscle) and NADPH-cytochrome reductase (liver). The old enzymes show severe losses in specific activity (30-60%)except for the phosphoglycerate kinases which are unchanged in this respect. All the enzymes show altered heat-sensitivity and where measured, immunological differences. None show differences in K_m or inhibitor effects. The young and old forms of nematode enolase and rat muscle phosphoglycerate kinase and glyceraldehyde-3-phosphate dehydrogenase showed readily detectable spectral differences.

As research into altered enzymes developed, evidence began to accumulate which was incompatible with the concept of errors. For example, isoelectric focusing showed that there was no difference in charge between young and old enzymes. This finding imposes a severe and unusual limitation on changes in sequence. Any amino acid substitutions would have to balance - base for base, acid for acid and neutral for neutral. Moreover, there could be no post-synthetic changes such as acylation, phosphorylation, dephosphorylation, sulfation, amidation or deamidation. Any of these would change the charge of the protein.

Enolase from young and old nematodes turned out to be a superior model for detailed study. Direct evidence that there were no covalent changes in the altered form of the enzyme was as follows: no methylation, no SH oxidation, no methionine sulfoxide formation, no loss of the C-terminal amino acid (no "cuts"), no change in charge - that is, no phosphorylation, deamidation, acylation, etc. The lack of plausible covalent changes supported the idea that the alteration in the old enzyme was due simply to conformational modifications. Direct and unequivocal evidence that there are no sequence changes or covalent post-synthetic changes was obtained from unfolding-refolding studies. Young and old forms of nematode enolase were each unfolded in 2M guanidine and then permitted to refold. The refolded young and old enzymes formed an identical product, similar but not quite the same as native old enolase. Immunotitration, heat sensitivity, inactivation by protease, C.D. and U.V. spectra support this conclusion (Sharma and Rothstein, 1980). These results proved unequivocally that altered enolase from old T. aceti possesses no errors.

Otherwise, the native forms could not refold to identical products.

As for mammalian enzymes, analytical studies of rat muscle phosphoglycerate kinase also have demonstrated that there are no "errors" or covalent changes. Young and old rat muscle PGK, respectively, were treated with trypsin, chymotrypsin and protease from _Staphylococcus aureus_. After treatment with each protease, and even a combination of proteases, the resulting peptides formed from young and old PGK were absolutely identical as determined by HPLC (Hardt and Rothstein, 1985). From these data, it is clear that there are no errors or post-synthetic covalent changes in old PGK. Therefore, old PGK must be a conformational isomer as is the case for nematode enolase. Such conformational change has been observed _in vitro_ for rat muscle 3-phosphoglyceraldehyde dehydrogenase (Gafni, 1983).

If altered enzymes consist of conformational isomers, how are they formed? We proposed 14 years ago that if protein turnover slowed with age, the enzymes would have an increased "dwell" time in the cells and the molecules would have time to become subtly denatured (Reis and Rothstein, 1974). Such altered molecules would accumulate, rather than be replaced. One can analogize with an enzyme solution mistakenly left overnight in the laboratory. In the morning, one might find that half of the activity had disappeared. Either half of the molecules were denatured and lost their activity, or all of the molecules were denatured and _each_ molecule lost one-half of its original activity. Most important, stable enzymes would be unaffected, thus explaining why we find that many enzymes are not altered during aging.

A probable example of this sort of conversion of enzyme molecules to inactive molecules has been shown in old rat kidney maltase. Using monoclonal antibodies, Reiss and Sacktor (1983) found that their old maltase consisted of a combination of normal enzyme and 57% inactive (old) molecules. The young preparation contained only 15% of inactive molecules. Two-dimensional gels indicated that there were no differences in the peptide composition of the two enzyme forms. Thus, the conclusion is inescapable that part of the young maltase is converted to inactive maltase in old rat kidney. Enzyme inactivation has also been reported in the lenses of the eye where there is no turnover of proteins (Dovrat et al. 1984).

From the above examples, it is clear that conformational changes in enzyme molecules can occur with age and that the altered molecules may or may not be active. Whether or not these changes are brought about by slowed protein turnover as earlier proposed, has not been proved unequivocally, but all of the available data fit the concept. In fact, in T. aceti, protein turnover slows dramatically with age, from a half life of 10 hours in 2-day old organisms, to over 250 hours at 28 days of age (Sharma et al. 1979). In mammals, studies have been less definitive, but all of the available evidence points to a slowing of protein turnover. Lavie et al. (1982) found that total soluble proteins of various crude fractions of mouse liver showed an increased half-life in aged animals. The rate of protein synthesis for rat liver (Ward, 1988) and cardiac proteins (Goldspink et al. 1986) has also been reported to decrease with age.

In addition to studies of individual enzymes from nematodes and rodents, several searches have been carried out for altered proteins on 2-dimensional gels. Thus, over 700 proteins from the free-living nematode Caenorhabditis elegans were compared at several ages. No significant qualitative differences were found (Johnson and McCaffrey, 1985). Fleming et al. (1985) made computerized comparisons of over 500 labeled proteins from young and old Drosophila by similar procedures. Again, there were no qualitative differences. Using cells in culture, Van Keuren et al. (1983) compared labeled proteins from early- and late-passage cells on 2-dimensional gels and observed only quantitative differences within a limit of 2.5% for abnormal proteins.

Another experimental approach to the idea of errors has been concerned with attempts to see if, with age, the fidelity of protein synthesis decreases in vitro. Such experiments, within the limits of detection, consistently failed to show a loss of fidelity in old preparations. In a study by Filion and Laughrea (1985), protein synthesis was claimed to be as accurate as the known values for the process in vivo. Meanwhile, evidence accumulated that many enzymes were not altered with age.

At present, about 25 such enzymes have been reported, including examples from the cytoplasm, mitochondria and lysosomes. The author is aware of about 10 more. The discovery of unaltered enzymes strikes a sharp blow against the idea of errors, since mistakes in transcription mediated by a faulty RNA polymerase would create errors in all proteins. The same reasoning applies to errors in translation. Clearly, if the large majority of enzymes is unaltered, the protein synthesizing machinery cannot be faulty.

The accuracy of DNA polymerase is directly involved with the Error Hypothesis in that the formation of faulty DNA could lead to errors in protein synthesis and thus initiate a catastrophe. However, current investigations of the enzyme show that there is no loss of fidelity with age. The accuracy of the enzyme does not change in old mice, either in short-lived Mus or long-lived Peromyscus. Recently, Silber et al. (1985) showed that regenerating liver of young and old mice showed no differences in fidelity. Both DNA polymerase-α and polymerase-β copies viral DNA with unchanged fidelity. The same was true of neuronal DNA polymerase-β from very old mice. The enzyme was error-prone, but there were no age-related differences. In addition, DNA polymerase in lymphocytes of old human subjects showed no loss of fidelity (Agarwal et al. 1978).

In brief, all of the experiments carried out with DNA polymerase have shown that there is no loss of fidelity with age. Obviously, there is little support for the idea of errors to be drawn from the area of DNA replication and much to be said against it.

Another possible source of errors is DNA repair. If repair is inadequate or faulty, errors could be generated that could lead to faulty proteins, loss of information of faulty regulation. Thus far, no examples of increased mis-repair due to aging have been found. In fact, for known types of damage, the rate of repair far exceeds the rate of damage. For that matter, it is generally accepted that neither excision repair nor strand break repair declines significantly with age.

In summary, it can be stated that there is overwhelming evidence against the existence of any kind of errors during aging. Investigations into enzyme structure have proved unequivocally that there are no errors in altered enzymes; the fidelity of protein synthesis remains unchanged with age; DNA has yet to be shown to contain errors. In fact, there is not a single piece of evidence which stands up to careful scrutiny and which provides direct evidence for errors in aged animals. It seems to be little noticed that Orgel himself long ago withdrew his support for the Error Hypothesis in 1973 (Orgel, 1973).

So, - how are proteins altered? The removal of "errors" from our lexicon does not mean that changes do not occur at the molecular level. Though most proteins are not altered with age, it has been established that in aged animals, some enzymes undergo conformational changes that affect their properties. We do not know how this happens. Various mechanisms of protein alteration can occur, but few have been demonstrated in an aging context. The mechanisms include the action of mixed function oxidases and oxidation and subsequent reduction of SH groups, topics to be addressed in this symposium. It should be pointed out that since protein turnover slows with age, altered proteins if they are formed, will tend to accumulate rather than be replaced. By the same token, slowed turnover implies changes in synthetic rates, thus perhaps altering effective levels of enzymes or hormones. On the other hand, surveys of enzyme levels have not uncovered dramatic changes with age. However, these determinations by no means eliminate the possibility that such changes occur for certain enzymes. Whatever the cause, altered regulation of various metabolic parameters is a persistent change observed during aging. Whether aging causes changes in regulation or regulation causes changes in aging is, of course, the key question. Certainly, it seems that many metabolic patterns are altered in dietarily restricted animals.

These individual events - changes in conformation or other kinds of alterations in individual proteins, be they enzymes, receptors or hormones, are a truer mark of aging than the all-encompassing idea of errors. It remains to identify the mechanisms involved, the proteins affected and the effects on the cell.

REFERENCES

Agarwal SS, Tuffner M, Loeb L (1978). DNA replication in human lymphocytes during aging. J Cell Physiol 96:235-244.

Dovrat A, Scharf J, Gershon G (1984). Glyceraldehyde-3-phosphate dehydrogenase activity in rat and human lenses and the fate of enzyme molecules in the aging lens. Mech Aging Dev 28:187-191.

Filian AM, Laughrea M (1985). Translation fidelity in the aging mammal: studies with an accurate in vitro system on aged rats. Mech Aging Dev 29:125-142.

Fleming JE, Quattrocki E, Latter G, Miguel J, Marcuson R, Zuckercandi E, Bensch KG (1985). Age-dependent changes in proteins of Drosophila melanogaster. Science 231:1157-1159.

Fry M, Loeb LA, Martin GM (1981). On the activity and fidelity of chromatin associated hepatic DNA polymerase in aging murine species. J Cell Physiol 106:435-444.

Gafni A (1983). Molecular origin of the molecular effects in glyceraldehyde-3-phosphate dehydrogenase. Biochim Biophys Acta 742:91-99.

Gershon H, Gershon D (1970). Detection of inactive molecules in aging of organisms. Nature 227:1214-1217.

Goldspink DF, Lewis SEM, Merry BJ (1986). Effects of aging and long term dietary intervention on protein turnover and growth of ventricular muscle in the rat heart. Cardiovasc Res 20:672-678.

Hardt H, Rothstein M (1985). Altered phosphoglycerate kinase from old rat muscle shows no change in primary structure. Biochim Biophys Acta 831:13-21.

Johnson TE, McCaffrey G (1985). Programmed aging or error catastrophe? An examination by two-dimensional polyacrylamide gel chromatography. Mech Aging Dev 30:285-297.

Lavie L, Reznick AZ, Gershon D (1982). Decreased protein and puromycinyl-peptide degradation in livers of senescent mice. Biochem J 202:47-51.

Orgel LE (1963). The maintenance of the accuracy of protein synthesis and its relevance to aging. Proc Natl Acad Sci USA 49:517-521.

Orgel LE (1970). The maintenance of the accuracy of protein synthesis and its relevance to aging: a correction. Proc Natl Acad Sci USA 67:1476.

Reiss U, Rothstein M (1974). Heat-labile isozymes of isocitrate lyase from aging Turbatrix aceti.

Biochem Biophys Res Comm 61:1012-1016.
Reiss U, Sacktor B (1983). Monoclonal antibodies to renal brush border membrane maltase: age-associated antigenic alterations. Proc Natl Acad Sci USA 80: 3255-3259.
Rothstein M (1982). "Biochemical Approaches to Aging". New York: Academic Press, pp 234-235.
Sharma HK, Prasanna HR, Lane RS, Rothstein M (1979). The effect of age on turnover in the free-living nematode Turbatrix aceti. Arch Biochem Biophys 194:275-282.
Sharma HK, Rothstein M (1980). Altered enolase in aged Turbatrix aceti results from conformational changes in the enzyme. Proc Natl Sci USA 77:5865-5868.
Silber JR, Fry M, Martin GM (1985). Fidelity of DNA polymerases isolated from regenerating rat liver chromatin of Mus musculus. J Biol Chem 260: 1304-1310.
Ward WF (1988). Enhancement by food restriction of liver protein synthesis in the aging Fischer 344 rat. J Gerontol 43:B50-53.
Van Keuran ML, Merril CR, Goldman D (1983). Protein variations associated with in vitro aging of human fibroblasts and quantitative limits on the error catastrophe hypothesis. J Gerontol 38:645-652.

The Role of Oxidative Modification in Cellular Protein Turnover and Aging

Pamela E. Starke-Reed

Laboratory of Biochemistry, National Heart, Lung, and Blood Institute, NIH, Bethesda, Maryland 20892

INTRODUCTION

Previous studies have demonstrated that many cellular enzymes accumulate as catalytically inactive or less active forms during aging (Rothstein, 1975; 1983). The mechanisms involved in this process have not been clearly defined. However, the possibility that some of the age-related alterations involve the oxidation of critical amino acid residues in enzymes is suggested by the demonstration that many of the enzymes which accumulate as inactive forms are highly susceptible to oxidative inactivation by various mixed function oxidation (MFO) systems *in vitro* (Fucci et al., 1983; Oliver et al., 1984). Moreover, oxidative damage is likely implicated in the turnover of enzymes and proteins. This view is supported by studies indicating that oxidatively inactivated enzymes exhibit increased susceptibility to known proteases, such as trypsin, subtilisin, and cathepsin D (Rivett, 1984) as well as by a new class of cytosolic proteases which selectively degrade oxidized proteins (Rivett, 1985a; 1985b). In addition, substrates and cofactors protect enzymes from oxidative inactivation suggesting that oxidation and proteolysis of the specific enzymes may be regulated by the availability of these metabolites (Fucci et al., 1983; Oliver et al., 1984; Oliver et al., 1981; Levine, 1984; Levine et al., 1984; Nakamura et al., 1985).

In order to investigate the physiological role of oxidative modification of proteins during aging *in vivo*, we have examined the levels of oxidatively modified proteins from young and old rats and young rats exposed to 100% oxygen. For these experiments hepatocytes were isolated from rats 3, 12, 20, and 26 months of age (Aging Model) or from rats exposed to 100% oxygen (Oxygen Toxicity Model) and the soluble protein fraction was prepared. Previous studies have indicated that MFO-mediated oxidative modification of proteins leads to the formation of a protein carbonyl derivative which can be detected by reactivity with 2,4-dinitrophenylhydrazine to form a stable hydrazone derivative (Levine, 1984). This property was used to quantitate the level of oxidized proteins in these extract preparations (Starke et al., 1987). The results show that the level of oxidized proteins increases with age from 3 to 26 months and also with exposure to 100% oxygen up to 48 hours followed by a sharp decrease between 48 and 54 hours. The specific activity of glutamine synthetase (GS) and glucose-6-phosphate dehydrogenase (G-6-PD) was decreased significantly during aging between 3 and 26 months and also during the 54-hour oxygen exposure. Antibody titration of GS and G-6-PD revealed that the amount of cross-reactive material decreased more slowly than the loss of enzyme activity during aging. However, during exposure to 100% oxygen the amount of immunologically reactive protein remained constant or increased with time of exposure to 100% oxygen until 48 hours and then markedly declined. These results show that loss of catalytic activity correlates with increased protein carbonyl content without loss of immunological cross reactivity. These studies suggest that inactive proteins accumulate in both experimental systems and are subsequently selectively degraded.

This report describes similar studies we have carried out with tissue from Lobund Aging Studies. We have examined the oxidative modification of soluble proteins in testes isolated from rats ages 6, 18, and 30 months. These animals were obtained from both the conventional full-fed or diet-restricted regimen and the germ-free, full-fed and diet-restricted regimen. Although the experimental models are quite different, the findings are compared with those obtained from isolated rat hepatocytes.

MATERIALS AND METHODS

Hepatocytes were isolated by the perfusion method of Seglen (1976) from rats of various ages or rats exposed to

100% oxygen previously described (Starke et al., 1987; Starke and Oliver, 1988). Rat testes were obtained from the Lobund Laboratory at the University of Notre Dame. The tissue samples were suspended in a buffer containing 137 mM NaCl, 4.6 mM KCl, 1.1 mM KH$_2$PO$_4$, 0.6 mM MgSO$_4$, 1.10 mM EDTA, 1.0 mM HEPES 0.5 µg/ml leupeptin, 40 µg/ml phenylmethylsulfonyl fluoride (PMSF), and 0.7 µg/ml pepstatin, pH 7.4. The cells were broken by sonication and insoluble debris was removed by centrifugation at 15,000 x g for 30 minutes. The supernatant fluid was retained and protein content was determined according to the Pierce BCA Protein Assay method. Protein carbonyl content was determined as previously described using 2,4-dinitrophenylhydrazine (DNPH) (Starke et al., 1987). The results represent the mean of duplicate determinations and are expressed as nanomoles of DNPH bound per milligram of protein. Glutamine synthetase activity was assayed by the γ-glutamyl transferase method of Rowe et al. (1970) as modified by Miller et al. (1978). Glucose-6-phosphate dehydrogenase activity was assayed according to the method of Bishop (17). DNPH was obtained from the Eastman Chemical Co. (Rochester, NY); leupeptin, pepstatin, and PMSF were obtained from Boehringer Mannheim (Indianapolis, IN); glutamine, NADP, glucose-6-phosphate, and ATP were obtained from Sigma Chemical Co. (St. Louis, MO). All other reagents were of the highest grade available.

RESULTS AND DISCUSSION

Oxidative inactivation of enzymes mediated by MFO systems in vivo or in vitro is accompanied by the generation of protein carbonyl derivatives which react readily with DNPH to form stable protein hydrazone derivatives yielding a characteristic difference spectrum with maximum absorbance at 365-375 nm. We have used this property to detect and quantitate the levels of oxidized proteins in cell-free extract preparations under conditions in which most of the lipids and small molecules are removed (Levine, 1984). Figure 1 shows the levels of protein carbonyl groups in the soluble fraction of hepatocytes isolated from rats aged 3, 12, 20 and 26 months (panel A) or rats exposed to 100% oxygen (panel B) and testes isolated from rats aged 6, 18, and 30 months full-fed or diet-restricted maintained in either a conventional or germ-free environment (panel C). In the hepatocyte samples protein oxidation increases with age as well as with exposure to 100% oxygen up to 48 hours followed by a sharp decline between 48 and 54 hours. In the testis samples from conventionally maintained rats the protein carbonyl levels increased with age in the full-fed animals (solid bars) but increased in the diet-restricted animals only up to 18 months (cross-hatched bars) then decreased between 18 and 30 months. The testis protein isolated from germ-free animals showed only a slight increase in protein carbonyl groups with age and there was little difference in the levels between the full-fed (right-hatched bars) and diet-restricted (left-hatched bars) rats.

Previous studies have indicated that a variety of kinases, dehydrogenases, synthetases, and other enzymes treated with enzymatic or non-enzymatic MFO systems are oxidatively inactivated by a site-directed mechanism (Fucci et al., 1983; Stadtman and Wittenberger, 1985; Levine, 1984). Other studies have demonstrated that oxidative inactivation of enzymes by MFO systems *in vitro* renders enzymes highly susceptible to proteolytic degradation by

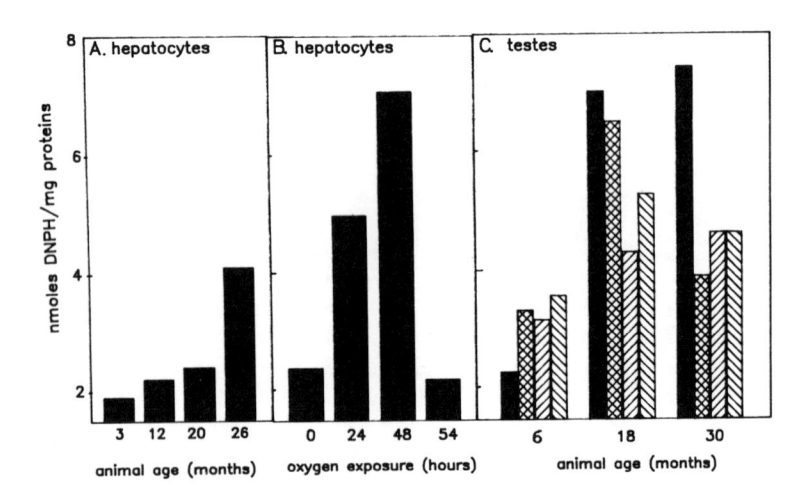

Figure 1. Levels of oxidized proteins in hepatocytes from rats of various ages (A) or from rats exposed to 100% oxygen (B) and in testes from rats conventionally grown full-fed (C; solid bars) or diet-restricted (C; cross-hatched bars) or germ-free rats full-fed (C; upward, right-hatched) and diet-restricted (C; upward, left-hatched).

subtilisin (Farber and Levine, 1982), cathepsin D (Rivett, 1985a), and a new class of cytosolic proteases which selectively degrade oxidatively modified proteins (Rivett, 1985a; 1985b). In order to determine whether the increase in protein oxidation observed in the hepatocytes and testes correlated with inactivation of specific cellular enzymes known to be susceptible to oxidative inactivation, we examined the activity of glutamine synthetase in these samples. Figure 2 shows the levels of glutamine synthetase in the hepatocytes (panels A and B) and testes (panel C) isolated at the various ages or during oxygen exposure at the indicated times. In hepatocytes there is a progressive loss of glutamine synthetase activity during aging and during exposure to 100% oxygen (solid lines). Antibody titration revealed that the amount of cross reacting material decreased more slowly with age and remained constant or increased with time of exposure to 100% oxygen until 48 hours and then markedly declined (data not shown).

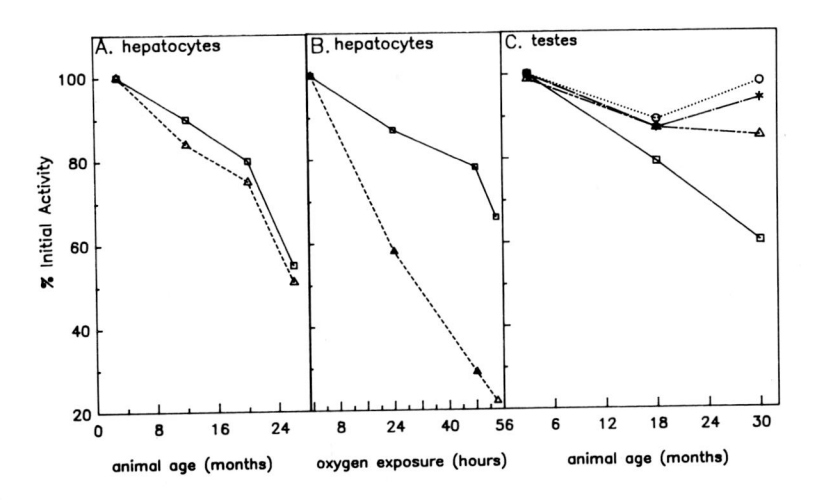

Figure 2. Inactivation of hepatocyte and testicular enzymes during aging or exposure to 100% oxygen. Glutamine synthetase (solid line) and glucose-6-phosphate dehydrogenase (dotted line) activities were assayed in hepatocytes isolated from rats of various ages (A) or rats exposed to 100% oxygen (B). Glutamine synthetase activity was assayed in testis extract from rats of various ages grown conventionally full-fed (solid line) or diet-restricted (dashed line) and germ-free rats full-fed (dotted dashed line) or diet-restricted (dotted line).

These findings suggest that the loss of activity is due to inactivation rather than to proteolysis of the enzyme. Furthermore, the fact that both the immunological cross reactivity and carbonyl content of the protein decreased between 48 and 54 hours of oxygen exposure is consistent with the selective degradation of oxidized proteins (Figure 1, panel B). Similar results were obtained with glucose-6-phosphate dehydrogenase activity (dashed lines). During oxygen exposure, however, G-6-PD was inactivated more rapidly than GS although the loss of cross reacting material was similar to that of GS. Glutamine synthetase activity was assayed in the four testis samples (Figure 2, panel C). The specific activity of GS in the full-fed rats maintained conventionally (solid line) decreased to 60% by 30 months. These results were comparable to the hepatocyte GS inactivation during aging. Glutamine synthetase activity in testes isolated from the conventionally maintained, diet-restricted rats (chain dashed line) was not inactivated to the same extent; only 13% of activity was lost. In contrast, the glutamine synthetase levels in the germ-free, full-fed (chain dotted) or diet-restricted (dotted line) rats remained relatively constant from ages 6 to 30 months. The GS enzyme inactivation patterns observed in testis tissue correlates inversely with generation of protein carbonyl groups suggesting that loss of enzyme specific activity is associated with increased oxidative modification as evidenced by increased carbonyl content. Other studies not shown here have demonstrated that proteolytic activity increased in the 48-54 hour interval during oxygen exposure and similar protease activities are deficient or defective in old animals.

These results demonstrate that oxidized proteins accumulate in vivo under a variety of experimental conditions but the mechanisms responsible for this accumulation may be quite different in different systems. Our studies with hepatocytes from animals subjected to treatment with 100% oxygen suggest that MFO activity may be responsible for the accumulation of oxidized proteins which are subsequently selectively degraded. In contrast, in hepatocytes obtained from animals of various ages, oxidized proteins may accumulate as a function of age because the proteases which selectively degrade these proteins are deficient or defective in old animals. In testis obtained from Lobund studies it is apparent that little change is seen in the level of oxidized proteins as a function of age except in the conventional, full-fed animals. In the three other regimens, reduced levels of oxidized proteins are observed indicating that diet restriction and germ-free environment had some protective effect which was apparently not additive and the mechanism of this protection is not yet known.

REFERENCES

Bishop C (1966). Assay of glucose-6-phosphate dehydrogenase (E.C.1.1.1.49) and 6-phosphogluconate dehydrogenase (E.C.1.1.1.43) in red cells. J Lab Clin Med 8:149-155.
Farber JM, Levine RL (1982). Oxidative modification of glutamine synthetase of E. coli enhances it susceptibility to proteolysis. Fed Proc 41:865.
Fucci L, Oliver CN, Coon MJ, Stadtman ER (1983). Inactivation reactions: Possible implications in protein turnover and aging. Proc Natl Acad Sci USA 80:1521-1525.
Levine RL (1984). Oxidative modification of glutamine synthetase. II. Characterization of the ascorbate model system. J Biol Chem 258:11828-11833.
Levine RL, Oliver CN, Fulks RM, Stadtman ER (1984). Turnover of bacterial glutamine synthetase: Oxidative inactivation precedes proteolysis. Proc Natl Acad Sci USA 78:2120-2124.
Miller RE, Hadenberg R, Gershman H (1978). Regulation of glutamine synthetase in cultured 3T3-L1 cells by insulin, hydrocortisone, and dibutyryl cyclic AMP. Proc Natl Acad Sci USA 75:1418-1422.
Nakamura K, Oliver CN, Stadtman ER (1985). Inactivation of glutamine synthetase by a purified rabbit cytochrome P450 system. Arch Biochem Biophys 240:319-329.
Oliver CN, Ahn B-W, Wittenberger ME, Stadtman ER (1985). Oxidative inactivation of enzymes; Implications in protein turnover and aging. In Ebashi S (ed): "Cellular Regulation and Malignant Growth," Berlin: Springer-Verlag, pp. 320-331.
Oliver CN, Levine RL, Stadtman ER (1981). Regulation of glutamine synthetase degradation. In Ornston LN, Silgar GS (eds): "Experiences in Biochemical Perceptions." New York: Academic Press, pp. 233-249.
Rivett AJ (1985). Preferential degradation of the oxidatively modified form of glutamine synthetase by intracellular mammalian proteases. J Biol Chem. 260:300-305.

Rivett AJ (1985). Purification of a liver alkaline protease which degrades oxidatively modified glutamine synthetase: Characterization as a high molecular weight cysteine proteinase. J Biol Chem 260:12600-12606.

Rivett AJ, Roseman JE, Oliver CN, Levine RL, Stadtman ER (1984). Covalent modification of proteins by mixed-function oxidation: Recognition by intracellular proteases. In Khairallah EA, Bond JS, Bird JD (eds): "Intracellular Protein Catabolism," New York: Alan R. Liss, pp. 317-328.

Rothstein M (1975). Aging and alteration of enzymes: A review. Mech Aging Dev 4:325-338.

Rothstein M (1983). Enzymes, enzyme alteration, and protein turnover. Rev Biol Res Aging 1:305-314.

Rowe WB, Remzio RA, Wellner VP, Meister A (1970). Glutamine synthetase (sheep brain). Methods Enzymol 17A:900-910.

Seglen PO (1976). Preparations of isolated rat liver cells. Methods Cell Biol 13:29-83.

Stadtman ER, Wittenberger Me (1985). Inactivation of Escherichia coli glutamine synthetase by xanthine oxidase, nicotinate hydroxylase, horse radish peroxidase, or glucose oxidase: Effects of ferredoxin, putidaredoxin, and menadione. Arch Biochem Biophy 238:379-387.

Starke PE, Oliver CN, Stadtman ER (1987). Modification of hepatic proteins in rats exposed to high oxygen concentration. FASEB J 1:36-39.

Starke-Reed PE, Oliver CN (1988). Oxidative modification of enzymes during aging and acute oxidative stress. In press.

AGE-RELATED MOLECULAR CHANGES IN SKELETAL MUSCLE

Ari Gafni and Khe-Ching M. Yuh

Institute of Gerontology and Department of Biological Chemistry, University of Michigan, Ann Arbor, Michigan 48109

INTRODUCTION

The occurrence of age-related modifications in functional and structural properties of a number of enzymes in skeletal muscle has been documented (Rothstein, 1983; Gafni, 1985). The study of these molecular aging effects and their mechanisms of development is of great interest since they can be characterized and understood in great detail. Moreover, the development of aging effects in certain enzymes may be used as a biomarker for biological aging or be compared with the alterations in enzymatic activities which are at the origin of certain metabolic diseases.

Dietary restriction has been shown to markedly extend the lifespan of laboratory rodents. This phenomenon is obviously of very great interest and potential importance; however, its origin is still not well understood. A comparative study of the aging process in enzymes in fullfed and diet-restricted rats is expected to deepen our understanding of the molecular basis of the life-prolonging effects of food restriction and to allow us to assess possible similarities in the development of the aging effects in the two groups of rats.

The following discussion addresses two enzyme systems from skeletal muscle in which age-related modifications have been identified and studied.

These are the glycolytic enzyme phosphoglycerate kinase (PGK) and the membrane-bound Ca^{2+}-dependent ATPase from the sarcoplasmic reticulum (SR-ATPase). These two enzymes will be utilized in our study of the effects of food restriction on molecular aging in muscle.

Aging Effects in PGK

Age-related modifications have been documented in phosphoglycerate kinase isolated from rat muscle (Sharma et al. 1980) liver (Hiremath, Rothstein 1982) brain (Sharma, Rothstein 1984) and heart (Zuniga, Gafni 1988). While the specific activity was found to be unaffected by age in all cases, several other properties of the enzyme did change significantly. In particular the thermal stability of old PGK was found to be markedly increased, compared with enzyme samples from young tissues, and this difference was frequently used to determine the status of PGK being studied. Detailed studies by Rothstein et al. (Rothstein 1985; Hardt, Rothstein 1985) revealed no covalent modifications in old

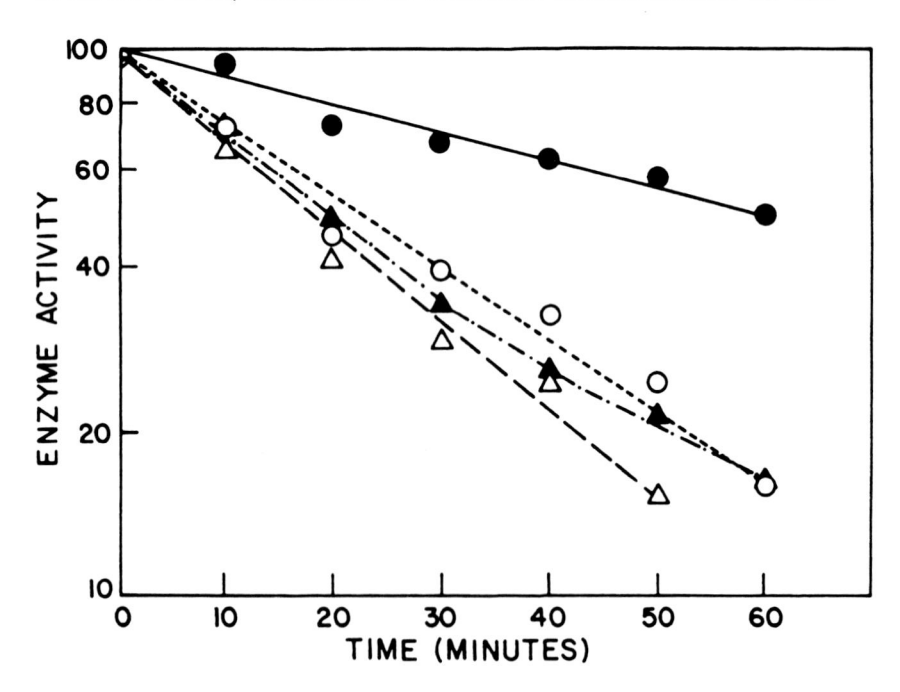

Figure 1. Heat inactivation of young (▲) and old (●) PGKs and of the corresponding unfolded-refolded forms of these enzymes (△ and ○) at $52°C$ in a 100 mM sodium phosphate buffer, pH 7.5/10 mM 2-mercaptoethanol/1mM EDTA.

PGK leading to the conclusion that the age-related effects in this enzyme reflect conformational modifications only. Strong support for this hypothesis was recently provided by our demonstration that old PGK may be efficiently converted to its young counterpart by extensive unfolding of the polypeptide chain followed by refolding under mild conditions to regain full enzymatic activity (Yuh, Gafni 1987; Zuniga, Gafni 1988). The results of such comparative unfolding-refolding experiments are presented in Figure 1 where the rate of heat inactivation of the various PGK samples is used as a sensitive probe for the enzymes status. Native old PGK is seen to be considerably more heat-stable than the young enzyme while unfolded-refolded old and young PGKs as well as untreated young PGK all display practically identical inactivation rates.

While the successful rejuvenation of old PGK reflects on the conformational nature of the age-related modifications in this enzyme, it does not shed light on their details or their location in the molecule. Studies aiming at these important aspects are currently underway.

Age-Related Effects in the SR-ATPase

A decrease, with aging, in the rate of Ca^{2+} uptake by cardiac SR-enriched fractions has been reported by several groups of investigators (Froehlich et al. 1978; Narayanan 1987). This age-related decline in calcium sequestration rate may be of major importance in explaining the prolonged relaxation time observed in hearts from old animals (Lakatta et al. 1975).

We have made a comparative study of the SR-ATPase in membrane fragments prepared from the skeletal muscles of Sprague-Dawley rats of several age-groups. A few properties of these preparations were found to become progressively modified with advancing age. Thus, the quantity of SR protein per gram of wet muscle tissue was markedly decreased in the old tissue as compared with young muscle. The capacity of SR vesicles for calcium ions was significantly reduced with age and the rate of inactivation of the ATPase upon incubation of intact SR vesicles at $37^\circ C$ was greatly increased at old age showing the old calcium pump system to be heat labile.

Alterations in the function of a membrane-bound enzyme

may reflect changes in the membrane, in the protein, or in each of these components. It is well established that many biological membranes become modified with age both in their phospholipid composition and in their degree of oxidation leading to marked changes in structure and functional properties. In the case of the SR-ATPase we now have strong evidence for the presence of age-related modifications in the protein molecule. Thus, when the ATPase is detached from the membrane and dissolved by the addition of Triton X-100, the degree of enzyme inactivation following a one hour incubation significantly differs between preparations

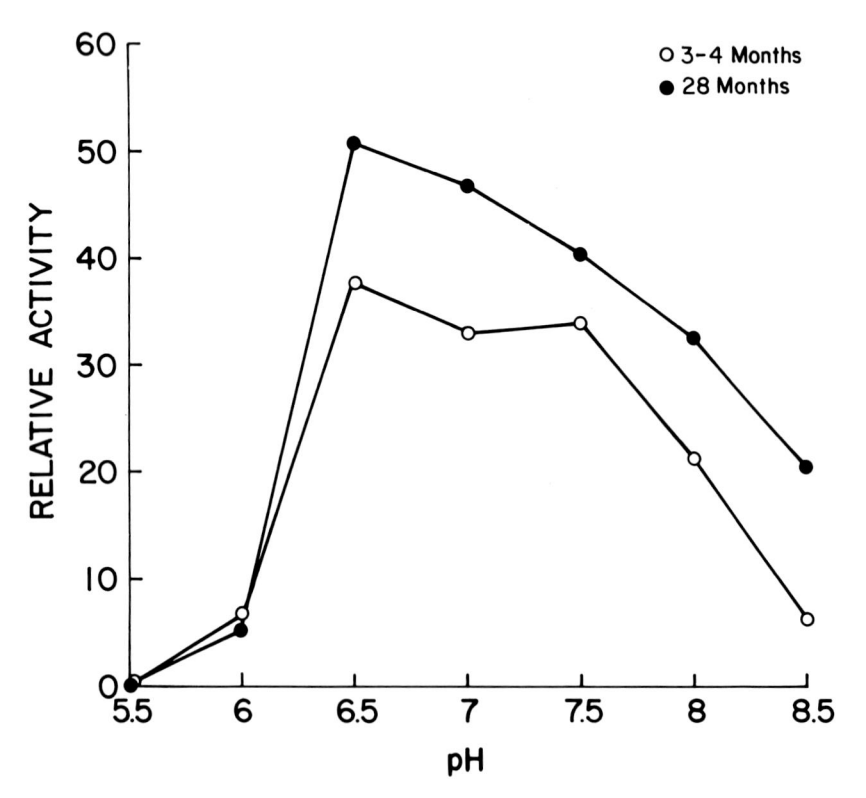

Figure 2. Effect of pH on the residual activity of SR-ATPase dissolved in 1% Triton X-100. The enzyme solutions from young (○) and old (●) rats were brought to the desired pH, incubated for 1 hour and their ATPase activity determined. Values given are relative to the activities of freshly dissolved samples of young and old ATPase.

from young and old rats. This is depicted in Figure 2 which shows the residual activities as a function of the pH. The enzyme stability is maximal at pH 6.5 and becomes negligible below pH 6.0 or above pH 8.5. Between these two values the stability of the old ATPase is consistently higher than that of the young enzyme.

The results attest to the presence of structural modifications in the old SR ATPase. These are currently being studied and how their development may be affected and modulated by dietary restriction is one major theme for our future work.

REFERENCES

Froehlich JP, Lakatta EG, Beard E, Spurgeon HA, Weisfeldt ML, Gerstenblith G (1978). Studies of sarcoplasmic reticulum function and contraction duration in young adult and aged rat myocardium. J Mol Cell cardiol 10: 427-438.

Gafni A (1985). Age-related modifications in a muscle enzyme. In Adelman RC, Dekker EE (eds.): "Modification of Proteins During Aging", New York: Alan R. Liss, PP 19-38.

Hardt H, Rothstein M (1985). Altered phosphoglycerate kinase from old rat muscle shows no change in primary structure. Biochim Biophys Acta 831:13-21.

Hiremath LS, Rothstein M (1982). The effect of aging on rat liver phosphoglycerate kinase and comparison with the muscle enzyme. Biochim Biophys Acta 705:200-209.

Lakatta EG, Gerstenblith G, Angell CS, Shock NW, Weisfeldt ML (1975). Prolonged contraction duration in aged myocardium. J Clin Invest 55:61-68.

Narayanan N (1987). Comparison of ATP dependent calcium transport and calcium activated ATPase activities of cardiac sarcoplasmic reticulum and sarcolemma from rats of various ages. Mech Ageing Dev 38: 127-143.

Rothstein M (1983). "Biochemical Approaches to Aging". London: Academic Press.

Rothstein M (1985). The alteration of enzymes in aging. In Adelman RC, Dekker EE (eds.): "Modification of Proteins During Aging", New York: Alan R. Liss, pp. 53-67.

Sharma HK, Prasanna HR, Rothstein M (1980). Altered phosphoglycerate kinase in aging rats. J Biol Chem 255:5043-5050.

Sharma HK, Rothstein M (1984). Altered brain phosphoglycerate kinase from aging rats. Mech Ageing Dev 25: 285-296.

Yuh KCM, Gafni A (1987). Reversal of age-related effects in rat muscle phosphoglycerate kinase. Proc Natl Acad Sci USA 84: 7458-7462.

Zuniga A, Gafni A (1988). Age-related modifications in rat cardiac phosphoglycerate kinase. Rejuvenation of the old enzyme by unfolding-refolding. Biochim Biophys Acta in the press.

DIETARY RESTRICTION POSTPONES THE AGE-DEPENDENT COMPROMISE OF MALE RAT LIVER MICROSOMAL MONOOXYGENASES

Douglas L. Schmucker and Rose K. Wang

Cell Biology & Aging Section, Veterans Administration Medical Center, the Department of Anatomy and the Liver Center, University of California, San Francisco, CA 94143

BACKGROUND

Dietary restriction remains the sole perturbation which significantly and universally influences life span (see Masoro, 1985 for a review). Specifically, reducing the caloric intake of animals extends life span by (a) altering certain physiological processes and (b) retarding age-associated pathologies. However, the mechanism(s) whereby dietary restriction affects these changes remains unresolved. Essentially all of the research in this arena has been performed using rodent models, i.e. inbred rats and mice. Furthermore, the majority of studies have been of a general nature and have not focused on particular cellular or subcellular systems. An exception to this pattern is the study demonstrating that age or dietary restriction-induced changes in the expression of certain hepatic proteins, i.e. a_{2u}-globulin, are elicited at the level of transcription (Richardson et al., 1987). Unfortunately this important study is tainted by the facts that: (a) a_{2u}-globulin synthesis is enhanced by androgens, (b) male Fischer 344 rats which are characterized by interstitial cell testicular tumors were used and (c) the specific function(s) of this protein remain unknown. Similar studies on well-characterized synthetic pathways and cell functions are essential to the dissection of the mechanisms whereby dietary restriction regulates the aging process.

The hepatic microsomal monooxygenases are well-characterized and the effects of aging on this system have been studied extensively, at least in inbred rodents (see Schmucker, 1985 for a review). Our laboratory has demonstrated that aging impairs the efficiency of this Phase I drug metabolizing pathway in this animal model, including (a) reduced levels and activities of specific monooxygenases and associated enzymes (Schmucker and Wang, 1980a; 1980b), (b) impaired phenobarbital-induction of monooxy- genases (Schmucker and Wang, 1981) and (c) the accumulation of catalytically inactive cytochrome P-450 reductase (Schmucker and Wang, 1983). In spite of the pool of catalytically inactive reductase, the livers of senescent rats contain an amount of enzyme similar to those of young animals which reflects the larger organ volume in the former age group (Schmucker et al., 1987). These data lend credence to the possibility that aging may not compromise in vivo hepatic Phase I drug-metabolism. Several investigators have reported significant sex-dependent shifts in the composition of the hepatic monooxygenases during aging in male rats (Kamataki et al., 1985; Kitagawa et al., 1985). However, the extrapolation of such rodent data to humans may have limited merit. Similar age and sex-related alterations in liver monooxygenases are not apparent in outbred rhesus monkeys (Maloney et al., 1986; Schmucker et al., 1987). The Lobund Study offered the opportunity to examine the impact of dietary restriction on a well-defined metabolic pathway, the hepatic monooxygenases, as well as to re-evaluate the effect(s) of aging and sex on these microsomal constituents in an alternative model, the Lobund-Wistar rat.

EXPERIMENTAL DESIGN

Conventional fed (CF) and restricted fed (CR) male Lobund-Wistar (L-W) rats were segregated into groups with average ages of 7 months (young adults), 18 months (mature), 26-36 months (old) and 37-42 months (very old; RF only). The maintenance and feeding regimens were similar to those described by Snyder and Wostmann (1987). Samples of intact liver tissue were obtained at necropsy, flash-frozen in liquid N_2 and shipped on dry ice to our laboratory. Microsomes were prepared and characterized as per Schmucker and Wang (1980), aliquoted and frozen at $-70°$ C until needed. The activity of NADPH cytochrome c (P-450) reductase was measured according to the method of Masters et al. (1967). The amount of this enzyme per mg of microsomal protein was estimated using rabbit anti-rat reductase polyclonal antibody (courtesy of Dr. Henry Strobel, University of Texas) in an ELISA. The total cytochromes P-450 content of the microsomes was estimated by (a) the CO-binding spectra (Omura and Sato, 1964) and (b) ELISA using a rabbit anti-rat cytochromes P-450 polyclonal antibody. The relative amounts of 16a (male specific) and 15b-hydroxylase (female specific) cytochromes P-450 were estimated via ELISA using mouse anti-rat monoclonal antibodies (courtesy of Dr. Edward Morgan, Emory University). All ELISA values were expressed as reciprocal titers. Microsomal protein content was measured by the method of Lowry et al. (1951). These data were subjected to statistical analyses (Student's t-test, ANOVA).

PRELIMINARY OBSERVATIONS

The specific activity of microsomal NADPH cytochrome c (P-450) reductase exhibited a marked decline (40%) between the young adult and old age CF groups. This loss of activity is consistent with that measured in similarly treated male Fischer 344 rats of approximately the same ages

(Schmucker and Wang, 1980). Interestingly, reductase activity remained relatively unchanged in the CR rats throughout the age span studied. Preliminary data from ELISA's using a polyclonal antibody directed against rat liver microsomal NADPH cytochrome P-450 reductase suggested higher enzyme titers in old and very old CR rats vs CF animals. In fact, the amount of enzyme present in the membranes of the oldest CR animals was similar to that of microsomes from young ad libitum fed rats. These values were not corrected for microsomal yield or liver volume.

The cytochromes P-450 content of CR rat microsomes measured via ELISA or shifts in the CO-binding spectra did not appear to exhibit an age-related decline until the animals were well into the fourth decade of life. Furthermore, the total cytochromes P-450 content was significantly increased in the membranes isolated from old CR animals. Unlike the male Fischer 344 rat, the immunoprecipitable cytochromes P-450 content of the hepatic microsomes in CF animals remained relatively unchanged (< 20% decline) during aging. Such interstrain variability was not unexpected since Gold and Widnell demonstrated marked differences in hepatic microsomal monooxygenases as a function of rat strain (Gold and Widnell, 1975). Analysis of the cytochromes P-450 content by shifts in the CO-binding spectra revealed a 45% decline between young and old CF rats. One possible explanation for the apparent disparity between these data and those generated via ELISA suggests that aging may impair the CO-binding capacity of these cytochromes while their immunoprecipitability remains unaffected. The relative membrane concentration of a male specific cytochrome P-450 isozyme, 16a-hydroxylase, in the CR animals reflected the age-related pattern exhibited by the total cytochromes P-450, i.e. the "feminization " of the male rat liver microsomal monooxygenase system does not occur in old CR animals (Kamataki et al., 1985). Shifts in the distribution pattern of female specific (15b-hydroxylase) during aging was unremarkable in either CF or CR rats.

We must caution that these data are very preliminary and await confirmation. However, they do afford some evidence that dietary restriction compensates for the documented age-related compromise of the hepatic microsomal monooxygenase system in ad libitum fed male rats. In essence, our results suggest that dietary restriction causes a distinct perturbation of a discrete and well-characterized enzyme system. In view of the recent report of enhanced synthesis of hepatic a_{2u}-globulin and its mRNA transcript in dietary restricted rats, our observations suggest that the hepatic monooxygenases may offer a model to examine the influence of dietary restriction at the transcriptional and translational levels in a well-characterized system (Richardson et al., 1987).

REFERENCES

Gold G and Widnell C (1975). Response of NADPH cytochrome c reductase and cytochrome P-450 in hepatic microsomes to treatment with phenobarbital-difference in rat strains. Biochem Pharmacol 24:2105-2106.

Kamataki T, Maeda K, Shimade M, Kitani K, Nagai T and Kato R (1985). Age-related alteration in the activities of drug-metabolizing enzymes and contents of sex-specifc forms of cytochrome P-450 in liver microsomes from male and female rats. J Pharmacol Exp Ther 233:222-228.

Kitagawa H, Fujita S, Suzuki T and Kitani K (1985). Disappearance of sex difference in rat liver drug metabolism in old age. Biochem Pharmacol 34:579-581.

Lowry OH, Rosebrough NJ, Farr AL and Randall RJ (1951). Protein measurement with the Folin phenol reagent. J. Biol Chem 193:265-275.

Maloney A, Schmucker DL, Vessey DA and Wang RK (1986). The effects of aging on the hepatic microsomal mixed function oxidase system of male and female monkeys. Hepatology 6:282-287.

Masoro EJ (1985). Nutrition and aging-a current assessment. J Nutr 27:98-101.

Masters BS, Williams C and Kamin H (1967). The preparation and properties of TPNH cytochrome c reductase from pig liver. Meth Enzymol 10:565-573.

Omura T and Sato R (1964). The carbon monoxide-binding pigment of liver microsomes. I. Evidence for its hemoprotein nature. J Biol Chem 239:2370-2378.

Richardson A, Butler JA, Rutherford MS, Semsei I, Gu M-Z, Fernandes G and Chiang W-H (1987). Effect of age and dietary restriction on the expression of a_{2u}-globulin. J Biol Chem 262:12821-12825.

Schmucker DL and Wang RK (1980a). Age-related changes in liver drug-metabolizing enzymes. Exp Gerontol 15:321-329.

Schmucker DL and Wang RK (1980b). Age-related changes in liver drug metabolism:structure vs function. Proc Soc Exp Biol & Med 165:178-197.

Schmucker DL and Wang RK (1981). Effects of aging and phenobarbital on the rat liver microsomal drug-metabolizing system. Mech Aging & Develop 15:189-202.

Schmucker DL and Wang RK (1983). Age-dependent alterations in rat liver microsomal NADPH cytochrome c (P-450) reductase:a qualitative and quantitative analysis. Mech Aging & Develop 21:137-156.

Schmucker DL, Vessey DA, Wang RK and Maloney AG (1987). Aging, sex and the liver microsomal drug-metabolizing system in rodents and primates. In Kitani K (ed): "Liver and Aging-1986," Amsterdam: Elsevier-North Holland, pp 3-14.

Schmucker DL and Wang RK (1987). Characterization of monkey liver microsomal NADPH cytochrome c (P-450) reductase as a function of aging. Drug Metab & Dispos 15:225-232.

Snyder DL and Wostmann BS (1987). Growth rate of male germfree Wistar rats fed ad libitum or restricted natural ingredient diet. Lab Animal Sci 37:320-325.

STIMULATION OF DNA CHAIN INITIATION BY A PROTEIN FACTOR (NPF-1) FROM RAT LIVER OF DIFFERENT AGES

Subhash Basu, Adrian Torres Rosado, Shigeo Takada, Satyajit Ray, Kamal Das, Isao Suzuki and Annie Pierre Seve.

Department of Chemistry, Biochem. Biophys. and Mol. Biol. Prog. University of Notre Dame, Notre Dame, IN 46556

INTRODUCTION

DNA replication in the eukaryotic system perhaps requires the concerted action of more than six enzymes: DNA polymerase-α (Kornberg,1988; Bollum and Potter, 1957; Hubscher, 1983), primase (Roth 1987; Conway and Lehman, 1982; Yagura et al., 1982; Tseng and Ahlem, 1983) RNase H (Crouch and Dirksen, 1982; Freeman-Wittig et al., 1986) DNA ligase (Sodertall and Lindahl, 1976), topoisomerase (Lin and Miller, 1981; Halligan et al, 1985, Liu et al., 1987) and helicase (Gilbert 1981; Wang, 1985) in addition to many unidentified nuclear proteins (Kalf et al., 1981 , Kawasaki et al., 1984; Takada et al., 1986). Control of these low molecular weight proteins may modulate DNA replication during development and aging. Aging of an organism is totally synchronized with the slowing down of DNA synthesis and the restriction of transcription and translational processes (Price et al., 1971; Turturro et al., 1986). However, very little is known about the requirements for initiation or the regulation of DNA replication in the eukaryotic system itself.

The T4 DNA chain initiation is catalyzed by the enzyme complex composed of DNA polymerase, primase and a few specific gene products (Alberts, 1984; Nossal, 1983). Initiation of SV-40 (Dodson et al., 1987) and PyV (DePamphilis and Bradley,1986) DNA replication requires specific interaction between the origin of the DNA replication sequence (Ori), T-antigen (T-ag) and permissive cell factors that consist of host cell DNA pol-α /primase.Ori is

composed of a DNA sequence which is strictly dedicated to
replication. However, in addition to T-antigens other
proteins which bind to promoter or enhancer sequences also
stimulate DNA synthesis. It has been realized that in the
eukaryotic system ori-cores of SV-40 (Dean et al., 1987) and
PyV (DePamphilis et al., 1987)) are different as to DNA
sequences or their upstream, near ori-core protein
interaction sequences. However, very little is known about
the roles of some of the nuclear nonhistone proteins in such
a role or in the stabilization of the DNA polα/primase
initiation complex such as 44, 62 and 45 gene products
(polymerase accessary proteins) in T-4 DNA replication
(Alberts, 1984; Nossal, 1983).

MATERIALS AND METHODS

Assay for DNA polymerase-α: DNA polymerase-α activity
(Weissbach et al., 1975; Chang et al., 1984) was measured by
the extent of incorporation of [3H]dCMP or [3H]TMP into
acid-insoluble activated calf thymus DNA (Simet, 1983). The
following components (in micromoles) were present in the
assay mixture in a total volume of 0.1 ml : activated calf
thymus, 20-40 µg; Tris-HCl, pH 8.0, 5.0; MgCl$_2$, 0.5; KCl,
2.0; DTT, 0.5; BSA, 50 µg, each unlabeled dNTP, 5 nmol; [3H]-
dCTP or [3H]dTTP (150-250 cpm/pmol); and enzyme (10 to 25 µg
protein). The mixtures were incubated for 45 min at 37°C.
The reaction was stopped by the addition of 50 µg BSA and
1.0 ml of ice-cold 10% TCA containing 100 mM sodium
pyrophosphate. After mild mixing on the vortex the mixture
was kept at 4°C for 30 min to complete the precipitation.
Precipitates were collected on GF/C glass fiber filters and
washed with 15 ml of ice-cold 5% TCA containing 50 mM sodium
pyrophosphate, 5 ml each of water, 95% ethanol, and acetone
for the removal of unincorporated dNTPs. After washing the
glass fiber filters were dried in a microwave oven and the
radioactivity incorporated into acid-precipitable material
was quantitated by liquid scintillation counting in a
Beckman model LS 3801 multichannel liquid scintillation
counter.

Assay for primase: To detect RNA chain-initiated DNA
synthesis, a coupled assay with DNA pol-α or Klenow (DNA
polymerase 78K) and synthetic template poly (dC) was used
(Takada et al., 1986) The following components (in micromoles) were present in the assay mixture in a total volume
of .05 ml : Poly (dC), 1.0 µg; Tris-HCl, pH 7.4, 2.5; MgCl$_2$,
0.4; DTT, 0.15; BSA, 5 µg; GTP or NTP (complementary to the
template sequence), 0.2; glycerol, 0.005 ml (v/v);

[3H]dGTP, 1.0-5.0 nmol (50 cpm/pmol); enzyme protein (0.1-10 μg). For the assay of free primase activity, Klenow (0.2 units) from United Biochemicals was used. The mixtures were incubated for 45 min at 37°C. The reaction mixtures were chilled at 4°C and spotted on DE-81 paper strips (3X2 cm). After air-drying, the strips (10-30 strips/250 ml) were washed 5 times with 3% Na_2HPO_4 (w/v), twice with water (250 ml), and once with 95% ethanol (250 ml). After washing, the strips were dried in a microwave oven and the amount of radioactivity incorporated was measured by liquid scintillation counting in a multichannel LS3801 Beckman scintillation counter.

Isolation of Free Primase by HIC: A novel hydrophobic interaction column chromatography has recently been developed in our laboratory for purification of glycosyl transferases (Basu et al., 1988) and has also been successfully used in the separation and purification of primase and DNA polymerase-α activities from PA-3 (rat prostate cells). The DNA polymerase-α / primase complex is initially purified from confluent cells (Campion, 1986) by the methods described previously (Takada et al., 1986; Simet et al., 1987;Ray et al.,1988). Treatment of PA-3 cell extract with ammonium sulfate (45%) dissociates (90%) the DNA pol-α/primase complex. The dissociated primase can then be purified by hydrophobic (octyl-sepharose) column chromatography. The detailed method will be published elsewhere.

Purification of NPF-1 from rat liver of different ages: Purified chromatin was prepared according to a modification of our previously published method (Bhattacharya et al., 1982) from the livers of 3- to 6-month-old Lobund Wistar rats. Freshly thawed rat liver (50 g)was homogenized with 4 volumes of Buffer A(Tris-HCl, pH 7.4, 15 mM; sucrose, 0.25M; $MgCl_2$, 2.5 mM; PEG(8K), 0.1%) and was centrifuged at 1500 X g for 15 min. The pellet (P-1) was resuspended in Buffer B (Tris-HCl) pH 7.4, 10 mM; sucrose, 2.3M; $MgCl_2$, 2.5 mM; KCl, 5mM; Triton X-100, 0.1%), rehomogenized and allowed to stir at 4°C for 15 min. The mixture was centrifuged in a Beckman L8M at 100,000 X g for 60 min. The pellet (P-2) was resuspended in 2 vol. of Buffer C-1 (Tris-HCl, pH 8.0, 10 mM; PEG (8K), 0.1%; sucrose, 0.25M; and $MgCl_2$, 2.5mM) and rehomogenized in a glass-teflon homogenizer. The mixture was centrifuged for 20 min at 15,000 X g. The pellet (P-3) was resuspended in 2 vols of Buffer C-1 and layered on top of a discontinuous sucrose density-(SDG) gradient (1.3M and 1.6M) containing Tris-HCl buffer, pH 8.0 (10.0mM) and PEG

(0.1%). The SDG was centrifuged for 2 hours at 100,000 X g and the chromatin pellet (P-4) was collected from the bottom of this gradient. The mixture of nonhistone chromatin proteins (NHCPs) was extracted from the pellet P-4 with increasing concentration of KCl (0.2 to 1.5M). The purified nuclear protein factor (NPF-1) which stimulates DNA pol-α/-primase-initiated DNA synthesis was purified by our previously described methods (Takada et al., 1986).

Assay for DNA-Unwinding Activity

DNA-unwinding activity was studied using S1 nuclease as described by Yoshida and Shimura (1984). Different amounts of NPF-1 (5-20 µg of protein) were incubated with 19 µg IMR-32 ds[3H]DNA (240 cpm/µg) in 10mM Tris-HCl(pH 7.0) at 37°C for 30 min. S1 nuclease was added to the mixture which was made 50mM NaOAc (pH 5.0) with 1 M NaOAc. After 30 min at 37°C, the nuclease reaction was stopped by cooling at 4°C with the simultaneous addition of ice-cold 10% (w/v) TCA. The mixtures were kept at 4°C for 30 min before centrifugation at 10,000 X g for 10 min at 4°C. The acid-soluble radioactivity was quantitated by spotting an aliquot of supernatant on GF/C filters followed by counting in a toluene scintillation system.

RESULTS AND DISCUSSION

From the combined KCl extract (0.75 to 1.5M) of rat liver chromatin, NPF-1 was purified by DE-23 column chromatography (Fig.1) and tested for purity by SDS-PAGE (Takada et al., 1986). In the presence of NPF-1 the incorporation of both [3H]dCMP in calf thymus DNA and [3H]dGMP in poly(dC) by IMR-32 DNA pol-α2 was stimulated 3-

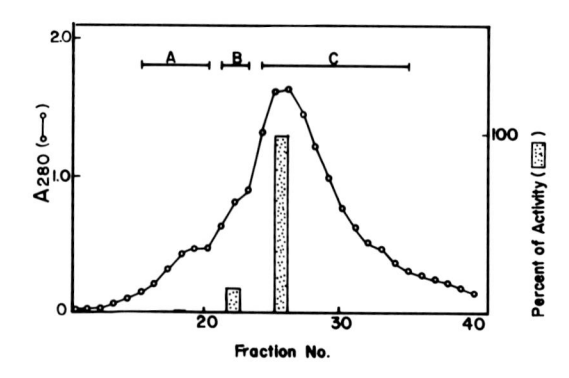

Figure 1. Purification of NPF-1 by DE-23 cellulose column chromatography. A linear gradient(0-2.0 M) KCl in 10 mM Tris-HCl(pH 8.0)/0.1% PEG/1 mM 2-ME(TPM buffer) was used.

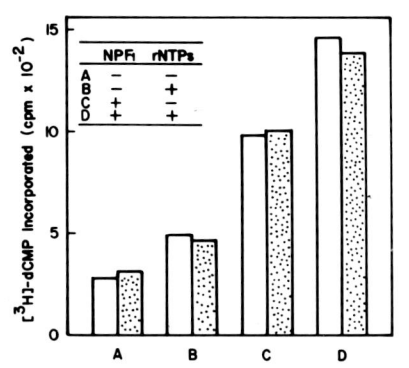

Figure 2. Effect of α-amanitin on ribonucleotide-stimulated DNA synthesis. Substrate was activated calf thymus DNA.

fold. The absence of RNA polymeraseII activity in the NPF-1 fraction was proved from the studies on the effect of α-amanitin on the reaction rates. As shown in Fig. 2, the rate of [3H]dCMP incorporation in ACT-DNA by IMR-32 DNA Pol-α /primase (+NPF-1) remained unchanged in the presence of α-amanitin (0.4mg/ml). The NPF-1 fraction obtained from 3- to

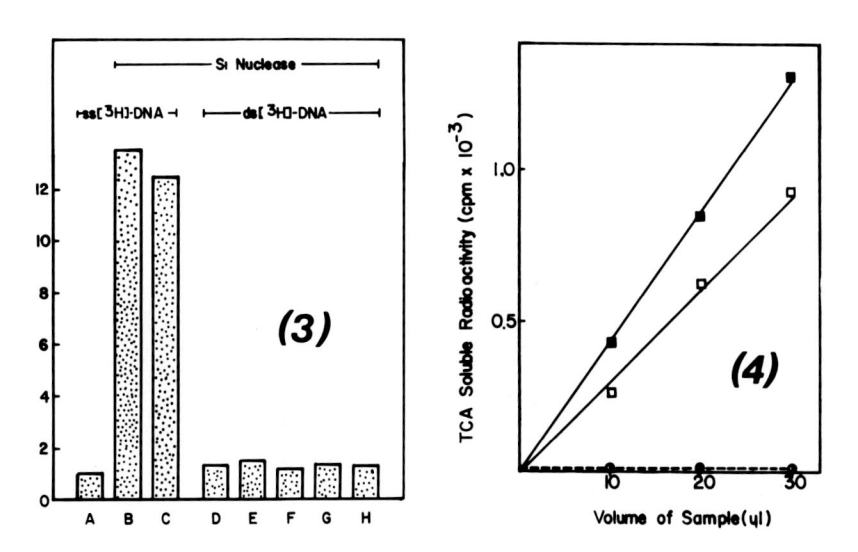

Figure 3. Test of DNA unwinding activity of NPF-1.

Figure 4. Test of DNase activity in NPF-1.

6- month-old rat liver contained neither unwinding activity (Fig. 3) nor nuclease activities (Fig. 4).

294 / Basu et al.

The effect of NHCP (0.5 to 1.5 M KCl extract of rat liver chromatin) and the purified NPF-1 on the stimulation of DNA synthesis has been tested in three different systems in our laboratory. The DNA pol-α/primase activities isolated from IMR-32 (Fig.5) and 9- to 11-day-old embryonic chicken brain (Fig.6) were also stimulated in the presence of NHCPs isolated from 3- to 6- month-old rat liver.

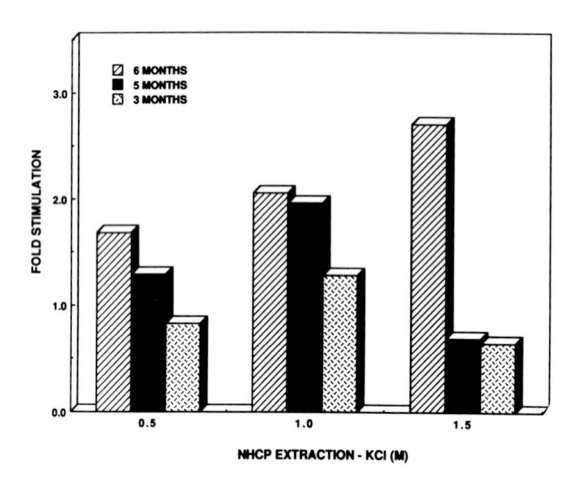

Figure 5. Stimulation of IMR-32 DNA pol-α/primase by rat liver NHCPs from various ages.

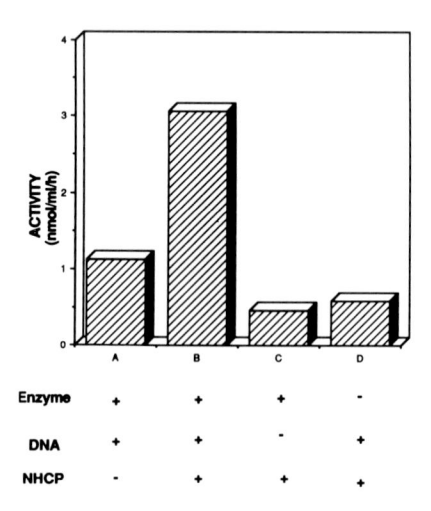

Figure 6. Effect of rat liver NHCP on DNA polymerase-α/-primase activities isolated from 11-day-old embryonic chicken brains. The enzyme fraction was isolated from 11-day-ECB by glycerol velocity gradient (10-30%:0.1 M KCl).

We have previously reported (Takada et al., 1986) that NPF-1 isolated from rat liver NHCP mixture stimulates RNA chain-initiated DNA synthesis and increases the number of transcripts rather than the chain length. We also suggested that perhaps NPF-1 stabilizes the DNA polα/primase initiation complex. However, the direct effect of any specific nuclear protein on the primase is not known. We have recently been able to obtain a highly purified DNA primase activity (Fig.7) from rat prostate PA-3 cells (Pollard and Luckert, 1975) by hydrophobic interaction chromatography. It appears that in the presence of 1.0 to 1.5 M KCl, there might be some factor in an extract of 5-month old rat liver that stimulates primase activity (Fig.8). Of current interest to us is whether this factor is similar to NPF-1. It is expected that these low molecular weight gene products are perhaps controlled during the aging process and thereby control DNA synthesis in the higher eukaryotes.

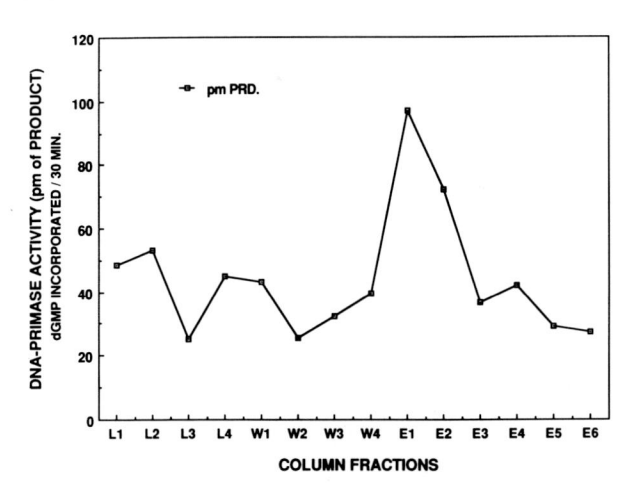

Figure 7. Purification of primase from PA-3 cells by octyl sepharose hydrophobic interaction column chromatography. The fraction was prepurified by DE-23 and blue agarose column chromatography.

The eukaryotic DNA polymerase holoenzyme is believed to be a multienzyme system similar to E.coli DNA polymerase III. Monoclonal antibodies which interact specifically with human KB-cell DNA pol-α (Hu et al., 1984; Wong et al., 1987) also bind or inhibit, in a similar way, the activities isolated from ECB, IMR-32 or PA-3 cells. A combined

Westernblot-immunoblot method(Ray et al.,1988) followed by activity determination after SDS-PAGE (Campion,1986) has been developed in our laboratory to determine molecular weights of the subunits present in the the DNA polα/primase complexes isolated from ECB,IMR-32 and PA-3 cells. Careful investigation of nuclear proteins may lead to better understanding of eukaryotic chain initiation. Recently a nuclear protein (PCNA) has been reported (Prelich and Stillman,1988) which is needed for SV-40 DNA replication.

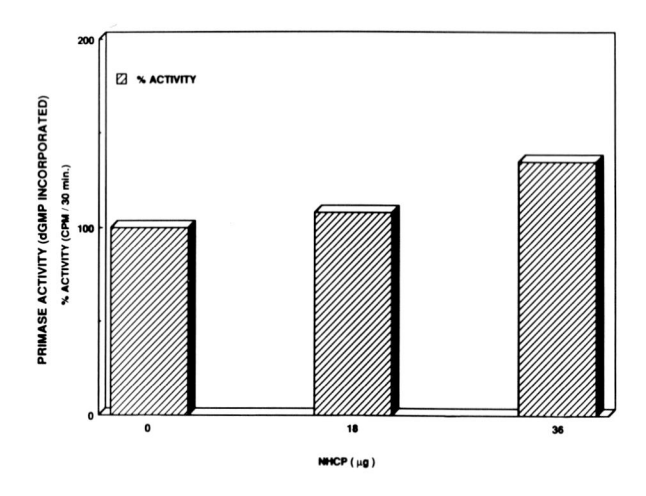

Figure 8. Stimulation of PA-3 primase by NHCPs isolated from 5-month-old rat liver. The combined fraction from 1.0 to 1.5 M KCl was used.

SUMMARY

DNA Polymerase-α/primase complexes have been isolated from human neuroblastoma IMR-32, embryonic chicken brains (ECB) and rat prostate tumor PA-3 cells. In the presence of $(NH_4)_2SO_4$ the major part (90%) of primase activity is released from the Pol-α/primase complex. A novel hydrophobic interaction column was used for purification of the primase from PA-3 cells. A nuclear protein factor (NPF-1) that stimulates DNA pol-α/primase activity has been purified from rat liver of various ages (3-6 months). The nuclear protein factor which only stimulates the primase activity is under investigation. The monoclonal antibodies (SJK 132-20 and 237-71) were used to detect DNA pol-α polypeptides from 11- to 19-day-old embryonic chicken brains.

ACKNOWLEDGEMENTS

This work was supported by NIH Grants NS-18005 and CA-14764. The authors wish to thank Dr. Manju Basu and Mr.Sujoy Ghosh for their help during preparation of this manuscript. We are also grateful to the Art Ehrman Cancer Fund and Retirement Research Foundation for providing some research funds during the course of this investigation for the development of the immunoblot assay.

REFERENCES

Alberts BM (1984). The DNA enzymology of protein machines. In Cold Spring Harbor symposium XLIX:1-12.
Basu S, Basu M, Das KK, Daussin F, Schaeper RJ, Banerjee P, Khan FA, Suzuki I (1988). Solubilized glycosyltransferases and biosynthesis in vitro of glycolipids. Biochimie (in Press).
Bhattacharya P, Basu S (1982). Probable involvement of a glycoconjugate in IMR-32 DNA synthesis: decrease of DNA polymerase 2 activity after tunicamycin treatment. Proc Natl Acad Sci USA 79:1488-1491.
Bollum FJ, Potter VR (1957). Thymidine incorporation into deoxyribonucleic acid of rat liver homogenates. JACS 79: 3603-3604.
Campion SR (1986). Characterization of DNA polymerase subunits from neuroblastoma cells in culture. Ph.D Thesis. University of Notre Dame, USA.
Chang LMS, Rafter E, Augl C, Bollum F (1984). Purification of a DNA polymerase - DNA primase complex from calf thymus glands. J Biol Chem 259:14679-14687.
Conway R, Lehman I (1982). A DNA primase activity associated with DNA polymerase from Drosophila melanogaster embryos. Proc Natl Acad Sci USA 79:2523-2527.
Cruch RJ, Dirksen ML (1982). In Lin SM, Roberts RJ (eds): "Nucleases", Cold Spring Harbor: Cold Spring Harbor Laboratory, pp 211.
Dean FB, Boroweic JA, Ishimi Y, Deb S, Tegtmeyer P, Hurwitz J (1987). Simian virus 40 large tumor antigen requires three core replication origin domains for DNA unwinding and replication in vitro. Proc Natl Acad Sci USA 84:8267-8271.
DePamphilis ML, Bradley MK (1986) Replication of simian and polyoma virus chromosomes. In Salzmann NP (ed): "The Viruses: Polyoma Viruses," New York: Alan R. Liss, pp 11-17.

DePamphilis ML, Decker RS, Yamaguchi M, Possenti R, Wirak DO, Perona R, Hassell JA (1987). Transcriptional elements and their role in activation of simian virus 40 and polyoma virus origins of replication. In DNA replication and recombination," New York: Alan R. Liss, pp 367-379.

Dodson M, Dean FB, Bullock P, Echols H, Hurwitz J (1987).Unwinding of duplex DNA from SV40 origin of replication by T-antigen. Science 238:964-967.

Freeman-Wittig MJ, Vinocour M, Lears R (1986) Differential effects of captan on DNA polymerase and ribonuclease H activities of avian myeloblastosis virus reverse transcriptase". Biochem 25:3050-3055.

Gellert M (1981). DNA Topoisomerases.Ann Rev Biochem 50: 879-910.

Halligan B, Edwards KA, Liu LF (1985). Purification and characterization of a type II DNA topoisomerase from bovine calf thymus. J Biol Chem 260:2475-2482.

Hu S, Wang T, Korn D (1984). DNA primase from KB cells: evidence for a novel model of catalysis by a highly purified primase/polymerase-α complex. J Biol Chem 259:2602-2609.

Hubscher U (1983). DNA polymerases in prokaryotes and eukaryotes: Mode of action and biological implications. Experientia 39:1-25.

Kalf GF, Metrione RM, Koszalka TR (1981). A protein stimulatory factor for DNA Polymerase-α in rat giant trophoblast cells. Biochem Biophys Res Commun 100:566-575.

Kawasaki K, Nagata K, Enomoto T, Hanaoka F, Yamada M-A (1984). Purification and characterization of a factor stimulating DNA polymerase-α activity from mouse FM3A cells. J Biochem 95:485-493.

Kornberg A (1988). DNA replication. J Biol Chem 263:1-4.

Liu LF, Miller KG (1981). Eukaryotic DNA topoisomerases: two forms of type$_I$ DNA topoisomerases from Hela cell nuclei. Proc Natl Acad Sci USA 78:3487-3491.

Liu Y, Liu LF, Li JJ, Wold MS, Kelly TJ (1987). The roles of DNA topoisomerases in SV40 DNA replication. "In DNA Replication and Recombination", New York: Alan R. Liss, pp 315-326.

Nossal NG (1983). Prokaryotic DNA replication systems. Ann Rev Biochem 53:581-615.

Pollard M, Luckert PH (1975). Transplantable metastasizing prostate adenocarcinomas in rats. J Natl Cancer Inst 51:643-649.

Prelich G, Stillman B (1988). Coordinated leading and lagging strand synthesis during SV40 DNA replication in

vitro requires PCNA. Cell 53:117-126.
Price GB, Modak SP, Makinodan T (1971) Age associated changes in the DNA of mouse tissues. Science 171:917.
Roth Y-F (1987). Eukaryotic Primase. Eur J Biochem 165:473-481.
Ray S, Torres-Rosado A, Suzuki I, Seve AP, Basu S (1988). Studies of DNA polymerase-α/primase complexes from embryonic chicken brain and rat prostate tumor cells. FASEB J 2:A1003.
Simet I (1983). Multiple forms of DNA polymerase- from embryonic chicken brain. Ph.D Thesis. University of Notre Dame, USA.
Simet I, Ray S, Basu S (1987). Resolution of DNA polymerase primase complex and primase free DNA polymerase from embryonic chicken brain. J Biosci 11:361-378.
Soderhall S, Lindahl T (1976). DNA ligases of eukaryotes. FEBS Letters 67:1-7.
Takada S, Torres-Rosado A, Ray S, Basu S (1986). Stimulation of human DNA polymerase-α and primase activities by a protein factor isolated from rat liver chromatin. Proc Natl Acad Sci US 83:9348-9352.
Tseng BY, Ahelm CN (1983). A DNA primase from mouse cells: purification and partial characterization. J Biol Chem 259:9845-9849.
Turturro A, Hart RW (1984) DNA repair mechanisms in aging. In "Comparative Pathobiology of Major Age-Related Diseases: Current Status and Research Frontiers," New York, pp 19-45.
Wang JC (1985). DNA topoisomerases. Ann Rev Biochem 54:665-697.
Weissbach A, Baltimore D, Bollum FJ, Gallo RC, Korn D (1975). Nomenclature of eukaryotic DNA polymerases. Science 190:401-402.
Wong SW, Wahl AF, Yuan PM, Arai N, Pearson BE, Arai KI, Korn D, Hunkapiller MW, Wang TS (1980). Human DNA polymerase-α gene expression is cell proliferation dependent and its structure is similar to both prokaryotic and eukaryotic replicative DNA polymerases. Embo J 47:37-47.
Yagura T, Kozu T, Seno T (1982). Mouse DNA replicase: DNA polymerase associated with a novel RNA polymerase activity to synthesize initiator RNA of strict size. J Biol Chem 257:11121-11127.
Yoshida M, Shimura K (1984). Unwinding of DNA by nonhistone chromosomal protein HMG (1+2) from pig thymus as determined with endonuclease. J Biochem 95:117-124.

EFFECT OF AGING AND DIETARY RESTRICTION ON THE EXPRESSION
OF α_{2u}-GLOBULIN IN TWO STRAINS OF RATS

Bo Wu, Craig C. Conrad and Arlan Richardson

Department of Chemistry
Illinois State University
Normal, Illinois 61761

INTRODUCTION

α_{2u}-Globulin is a major urinary protein of male rats. It is a low molecular weight protein and immunochemically distinct from other proteins. Initially, Roy et al. (1966) described a urinary protein that had an electrophoretic mobility similar to the serum protein α_2-globulin. However, it was distinctly different from the serum α_2-globulin, and it was renamed α_{2u}-globulin, the "u" indicating urine. α_{2u}-Globulin is synthesized and secreted by the hepatic parenchyma cells and is rapidly filtered through the kidneys into the urine (Roy et al., 1966; Roy and Neuhaus, 1966; Roy and Raber, 1972). Because of its rapid filtration rate, α_{2u}-globulin makes up only 0.5% of the serum proteins.

α_{2u}-Globulin actually consists of a family of low-molecular-weight proteins. They are resolved into two distinct molecular forms (M_r 18,800 and M_r 18,100) by sodium dodecylsulfate (SDS) polyacrylamide gel electrophoresis. The ratio of the two molecular forms of α_{2u}-globulin varies in different strains of rats (Chatterjee, 1982). Five major and two minor isoelectric variants of α_{2u}-globulin within the total hepatic proteins have been identified by 2-dimensional polyacrylamide gel electrophoresis (Roy et al., 1983). In addition, a small amount (~3%) is post-translationally modified to contain a 2500-dalton carbohydrate residue (Chatterjee et al., 1982). The messenger RNA for α_{2u}-globulin constitutes 1 to 2% of the total hepatic mRNA. The mRNA for α_{2u}-globulin contains approximately 1200 nucleotide residues with 300 to 400 nucleotides in the non-coding sequences and about 175 adenosine residues in the 3'-poly(A) segment (Deshpande et al., 1979; Chatterjee and Roy, 1980)

α_{2u}-Globulin is an androgen-dependent protein. It appears at puberty (approximately 6 weeks of age). Castration of mature male rats results in a sharp decrease in the urinary level of α_{2u}-globulin. It is absent in female and immature rats. Treatment of castrated female rats with either testosterone or 5α-dihydrotestosterone results in the appearance of α_{2u}-globulin (Roy and Neuhaus, 1967). On the other hand, estrogenic hormones can completely inhibit the urinary output of α_{2u}-globulin in male rats (Roy et al., 1975). More recently, it was found that growth hormone would affect α_{2u}-globulin synthesis independent of androgens (Murty et al., 1987).

α_{2u}-Globulin is of interest because it is under multihormonal control. Treatment of hypophysectomized male rats with testosterone, along with a single pituitary hormone, failed to reverse the effect of hypophysectomy (Roy, 1973a). Complete recovery from the effect of hypophysectomy required treatment with a combination of four hormones: testosterone, corticosterone, thyroxine, and growth hormone (Roy, 1973). In addition, it has been shown that insulin plays an important role in the regulation of α_{2u}-globulin. Alloxan induced diabetes results in a sharp decrease in the daily urinary excretion of α_{2u}-globulin. α_{2u}-Globulin in diabetic rats can be raised to nearly normal levels and maintained with insulin supplementation (Roy and Lenorard, 1973).

Steroid hormones, insulin, and growth hormone regulate the hepatic synthesis of α_{2u}-globulin through changes in the hepatic concentration of α_{2u}-globulin mRNA (Sippel et al., 1975; Roy et al., 1977; Roy et al., 1980; Roy and Dowbenko, 1977; Roy et al., 1982). It is of interest that the androgen-mediated increase in the hepatic concentration of α_{2u}-globulin mRNA may only be part of the inductive response. Hormone-dependent regulation of α_{2u}-globulin secretion may constitute another regulatory component (Murty et al., 1987a). The effect of thyroxine on α_{2u}-globulin synthesis is correlated to changes in the hepatic concentration of α_{2u}-globulin mRNA. However the effect of thyroxine on α_{2u}-globulin mRNA levels is indirectly mediated through growth hormone. Thyroxine also seems to directly influence the translation efficiency of α_{2u}-globulin mRNA (Kurtz et al., 1976; Chatterjee et al., 1983).

In 1981, Roy et al. reported that the levels and synthesis of α_{2u}-globulin decreased dramatically with increasing age in Fischer F344 rats. Therefore, Roy proposed that α_{2u}-globulin was a senescence biomarker (Chatterjee et al., 1981). In this article, we will describe the effect of aging and dietary restriction on the expression of α_{2u}-globulin in male Fischer F344 rats and Lobund/Wistar rats. Dietary restriction (underfeeding not malnutrition) is the only experimental manipulation known to increase the longevity of mammals (Richardson, 1985; Masoro, 1985). Dietary restriction is believed to increase longevity by retarding the aging process, e.g., dietary restriction (a) increases the maximum as well as the mean survival, (b) retards the incidence and severity of a wide variety of diseases, (c) retards the decline in a variety of physiological processes, and (d) retards the decline in the function of a variety of organs/tissues (Masoro, 1985).

RESULTS AND DISCUSSION

Our laboratory has studied the effect of aging on the expression of α_{2u}-globulin in male Fischer F344 rats (Richardson et al., 1987). Figure 1 summarizes the results of our previous study. The level of α_{2u}-globulin in the urine decreased dramatically with increasing age eventhough the total amount of protein in the urine increased. The age-related increase in proteinuria is well documented in laboratory rodents (Ricketts et al., 1985). Figure 1 shows that the age-related decline in the levels of α_{2u}-globulin in the urine was correlated to an age-related decline in the ability of isolated hepatocytes to synthesize α_{2u}-globulin. Subsequently, we found that the age-related decline in α_{2u}-globulin synthesis was paralleled by a decline in α_{2u}-globulin mRNA levels and the transcription of the α_{2u}-globulin genes. Thus, it appears that the age-related decline in α_{2u}-globulin levels in the urine is due to a decline in the transcription of α_{2u}-globulin.

Subsequently, our laboratory studied the effect of dietary restriction on the age-related decline in α_{2u}-globulin expression(Richardson et al., 1987), and these results are summarized in Figure 2. In this study, male

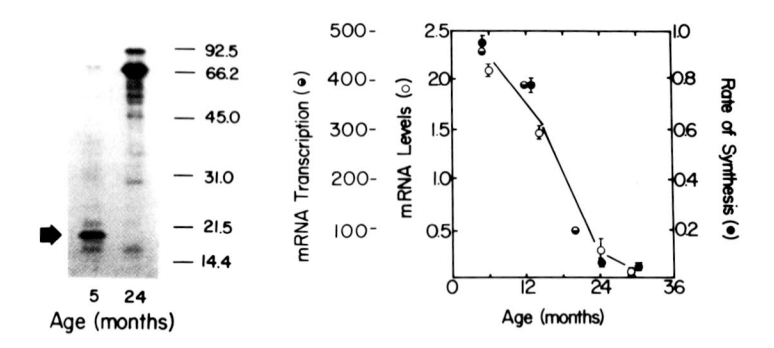

Figure 1. Effect of aging on the expression of α_{2u}-globulin (data taken from Richardson, et al., 1987). The separation of proteins from the urine of male Fischer F344 rats by SDS-gel electrophoresis is shown on the left. The arrow indicates the migration of α_{2u}-globulin. The synthesis of α_{2u}-globulin , α_{2u}-globulin mRNA levels , and the transcription of α_{2u}-globulin genes are shown in the graph on the right.

Fischer F344 rats were placed on two dietary regimens at 6 weeks of age. One group was fed ad libitum and the second group was fed 60% of the diet consumed by the rats fed ad libitum (restricted). Yu et al. (1982) showed that this restriction procedure increased the survival of male Fischer F344 rats substantially (Yu et al., 1982). The expression of α_{2u}-globulin was measured when these rats reached 18 months of age. Figure 2 shows that the level of α_{2u}-globulin in the urine of rats fed the restricted diet was higher than that found in the urine of rats fed ad libitum. In addition, it is apparent that the level of protein in the urine of the restricted rats was much lower than that of the rats fed ad libitum. Previous studies have shown that dietary restriction reduces that age-related increase in proteinuria (Ricketts et al., 1985). Figure 2 also shows that the increase in α_{2u}-globulin is due to an increase in the expression of α_{2u}-globulin. An increase in the synthesis, mRNA levels, and transcription of α_{2u}-globulin are observed with dietary restriction. Thus, dietary restriction prevented the age-related decline in the expression of α_{2u}-globulin in male Fischer F344 rats at the level of transcription.

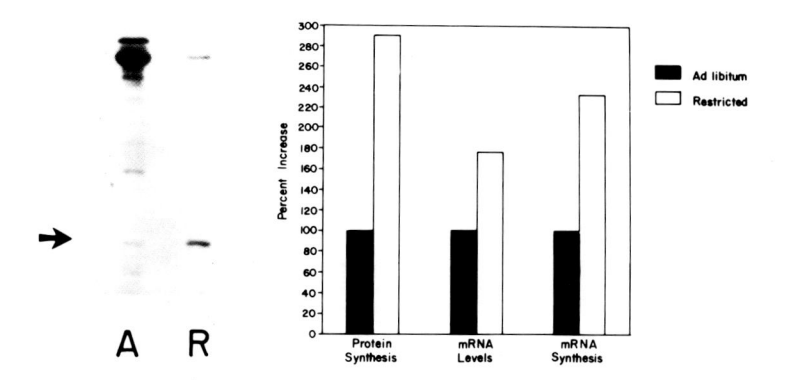

Figure 2. The effect of dietary restriction on the expression of α_{2u}-globulin (data taken from Richardson et al., 1987). The expression of α_{2u}-globulin was measured at 18 months of age in male Fischer F344 rats fed ad libitum (A) or the restricted diet (R). The figure on the left shows the separation of proteins in the urine by SDS-gel electrophoresis (the arrow indicates the migration of α_{2u}-globulin). The graph on the right shows the synthesis, mRNA levels, transcription of α_{2u}-globulin in the rats fed ad libitum and the restricted diet.

Because of our studies with male Fischer F344 rats, we were interested in studying the effect of aging and dietary restriction on the expression of α_{2u}-globulin in male Lobund/Wistar rats from the animal colony maintained at the University of Notre Dame. In these experiments, urine was collected over a 24-hour period from 7-, 18-, and 30-month-old rats maintained under conventional, barrier conditions or germ-free conditions. The rats were fed ad libitum or a restricted diet beginning at 8 weeks of age. Figure 3 shows the separation of proteins in the urine of the Lobund/Wistar rats maintained under conventional conditions. It is apparent that the levels of α_{2u}-globulin in the urine of these rats changed very little between 7 and 30 months of age. This is in marked contrast to what we had observed in male Fischer F344 rats (Figure 1). In addition, dietary restriction had very little effect upon the level of α_{2u}-globulin in the urine of the Lobund/Wistar rats. However, we did observe that dietary restriction prevented the

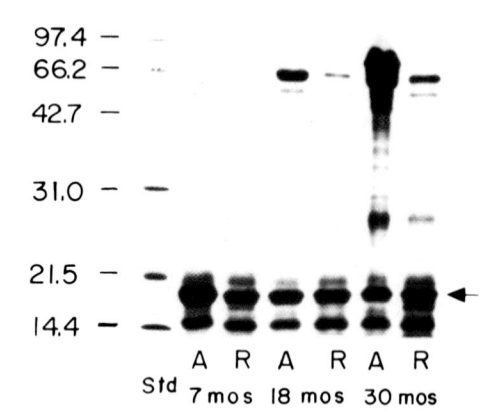

```
97.4 —
66.2 —
42.7 —

31.0 —

21.5 —
14.4 —
        A  R  A  R  A  R
   Std  7 mos  18 mos  30 mos
```

Figure 3. Levels of α_{2u}-globulin in the urine of male Lobund/Wistar rats maintained under conventional conditions. Urine was collected from rats fed <u>ad libitum</u> (A) or a restricted diet (R) at the ages shown. Urine was collected over a 24 hour period and was pooled from four rats in each group. The migration of protein standards is shown, and the arrow indicates the migration of α_{2u}-globulin.

increase in proteinuria that occurred in 30-month-old rats fed <u>ad libitum</u>. Figure 4 shows that similar results were obtained with male Lobund/Wistar rats maintained under germ-free conditions.

The results of our preliminary experiments with male Lobund/Wistar rats indicate that the urinary levels of α_{2u}-globulin change very little with age or dietary restriction. This is in contrast to what we observed in male Fischer F344 rats (Figure 1 and 2). The level of α_{2u}-globulin decreased markedly with age, and dietary restriction maintained higher levels of α_{2u}-globulin. The reason for the differences in α_{2u}-globulin expression in these two strains of rats is not known. It is possible that testicular tumors might contribute to the decline in α_{2u}-globulin expression through changes in androgen levels. The incidence of interstital testicular tumors increase markedly with age in Fischer F344 rats (Maeda et al., 1985). It is also possible that changes in α_{2u}-globulin expression might be observed in older Lobund/Wistar rats because the

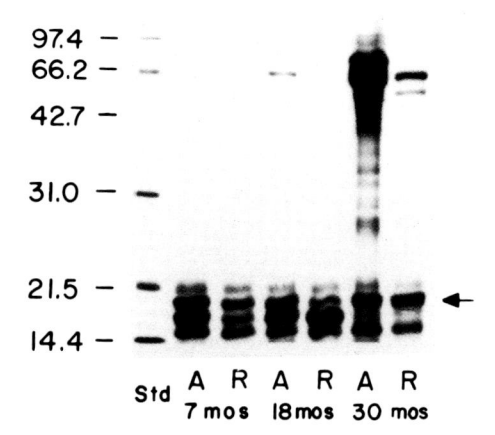

```
97.4 —
66.2 —
42.7 —
31.0 —
21.5 —
14.4 —
     Std  A  R  A  R  A  R
         7 mos  18 mos  30 mos
```

Figure 4. Level of α_{2u}-globulin in the urine of male Lobund/Wistar rats maintained under germ-free conditions. The data are presented as described in Figure 3.

Lobund/Wistar colony of rats maintained at the University of Notre Dame has a survival that is six to seven months longer than that normally reported for male Fischer F344 rats.

ACKNOWLEDGMENTS

These experiments were supported by Grant AG 01548 from the National Institutes of Health.

REFERENCES

Chaterjee B, Roy AK (1980). Messenger RNA for α_{2u}-globulin of rat liver. J Biol Chem 255:11607-11612.

Chatterjee B, Nath TS, Roy AK (1981). Differential regulation of the messenger RNA for three major senescence marker proteins in male rat liver. J Biol Chem 256:5939-5941.

Chatterjee B, Motwani NM, and Roy AK (1982). Synthesis and processing of the dimorphic forms of rat α_{2u}-globulin. Biochim Biophys Acta 698:22-28.

Chatterjee B, Demyan WF, Roy AK (1983). Interacting role of thyroxine and growth hormone in the hepatic synthesis of

α_{2u}-globulin and its messenger RNA. J Biol Chem 258:688-692.
Deshpande AK, Chatterjee B, Roy AK (1979). Translation and stability of rat liver messenger RNA for α_{2u}-globulin in Xenopus oocyte: the role of terminal poly(A). J Biol Chem 254:8937-8942.
Kurtz DT, Sippel AE, Fregelson P (1976). The effect of thyroid hormones on the level of the hepatic mRNA for α_{2u}-globulin. Biochemistry 15:1031-1036.
Maeda H, Gleiser CA, Masoro EJ, Murata I, McMahan CA, Yu BP (1985). Nutritional influences on aging of Fischer 344 rats: II. Pathology. J Gerontol 40:671-688.
Masoro EJ (1985) Nutrition and aging--a current assessment J Nutr 115:842-848.
Murty CVR, Rao KVS, Chung KW, Roy AK (1987). Independent regulatory influence of growth hormone on the hepatic synthesis of α_{2u}-globulin. Endocrinology 121:1819-1823.
Murty CVR, Rao KVS, Roy AK (1987a). Rapid androgenic stimulation of α_{2u}-globulin synthesis in the perfused rat liver. Endocrinology 121:1814-1818.
Richardson A (1985). The effect of age and nutrition on protein synthesis by cells and tissues from mammals. In Watson, RR (ed.): "Handbook of nutrition in the aged," Boca Raton, FL: CRC Press, pp 31-47.
Richardson A, Butler JA, Rutherford MS, Semsei I, Gu MZ, Fernandes G, Chiang WH (1987). Effect of age and dietary restriction on the expression of α_{2u}-globulin. J Biol Chem 262:12821-12824.
Richetts WG, Birchenall-Sparks MC, Hardwick JP, Richardson A (1985). Effect of age and dietary restriction on protein synthesis by isolated kidney cells. J Cell Physiol 125:492-498
Roy AK, Neuhaus OW (1966). Proof of the hepatic synthesis of a sex-dependent protein an the rat. Biochim Biophys Acta 127:82-87.
Roy AK, Neuhaus OW, Harmison CR (1966). Preparation and characterization of a sex-dependent rat urinary protein. Biochim Biophys Acta 127:72-81.
Roy AK, Neuhaus OW (1967). Androgenic control of a sex-dependent protein in the rat. Nature 214:618-620.
Roy AK, Raber DL (1972). Immunofluorescent localization of α_{2u}-globulin in the hepatic and renal tissues of rat. J Histochem Cytochem 20:89-96.
Roy AK (1973). Androgenic induction of α_{2u}-globulin in rats: androgen insensitivity in prepubertal animals. Endocrinology 92:57-61.

Roy AK (1973a). Androgen-dependent synthesis of α_{2u}-globulin in the rat: role of the pituitary gland. J Endocrinol 56:295-301.

Roy AK, Lenonard S (1973). Androgen-dependent synthesis of α_{2u}-globulin in diabetic rats: the role of insulin. J Endocrinol 57:327-328.

Roy AK, McMinn DM, Biswas NM (1975). Estrogenic inhibition of the hepatic synthesis of α_{2u}-globulin in the rat. Endocrinology 97:1501-1508.

Roy AK, Dowbenko DJ (1977). Role of growth hormone in the multi hormonal regulation of mRNA for α_{2u}-globulin in the liver of hypophysctomized rats. Biochemistry 16:3918-3922.

Roy AK, Dowbenko DJ, Schiop MJ (1977) Studies on the mode of estrogenic inhibition of hepatic synthesis of α_{2u}-globulin and its corresponding messenger ribonucleic acid in rat liver. Biochem J 164:91-97.

Roy AK, Chatterjee B, Prasad MSK, Unakar NJ (1980). Role of insulin in the regulation of hepatic messenger RNA for α_{2u}-globulin in diabetic rats. J Biol Chem 255:11613-11618.

Roy AK, Chatterjee B, Demyan WF, Nath TS, Motwani NM (1982). Pretranslational regulation of α_{2u}-globulin in rat liver by growth hormone. J Biol Chem 257:7834-7838.

Roy AK, Nath TS, Motwani NM, Chatterjee B (1983). Age dependent regulation of the polymorphic forms of α_{2u}-globulin. J Biol Chem 258:10123-10127.

Sippel AE, Feiglson P, Roy AK (1975). Hormonal regulation of the hepatic messenger RNA levels for α_{2u}-globulin Biochemistry 14:825-829.

Yu BP, Masoro EJ, Murata I, Bertrand HA, Lynd TT (1982). Life span study of SPF Fischer 344 male rats fed <u>ad libitum</u> of restricted diets: longevity, growth, lean body mass and disease. J Gerontol 37:130-141.

Index

Abdominal palpation, 53
Acini
   constricted, 56–57
   mucous, 80–81
Actinomyces sp., 136
Acyclation, 261
Addison's disease, 196
   corticosteroid substitution for, 197
Adenocarcinoma prostate, 53, 55–56, 130
   metastatic, 58–59
   moderately differentiated, 55
Adenofibroma, mammary, 58, 144
Adenohypophysial
   cells, 185, 192, 194, 197
   changes in conventional, germfree, and food-restricted aging Lobund–Wistar rats, 181–190
   hormones, 185, 192
Adenoma, 198
   ACTH-secreting, 197
   benign, 39
   cells, 185–186
   FSH-producing, 185, 187–188
   gonadotroph, 183–185, 187–188, 196–198
   human null cell, 198
   lactotroph, 184–185, 187–188, 192
   LH-producing, 185, 187–188
   lung, 58
   parathyroid, 58
   pituitary, 182–186, 192, 197–198
   prolactin-producing, 184–185, 187–188, 192
   thyrotroph, 184, 196
   TSH-producing, 184–185, 187
Adenosine triphosphate (ATP), 100
Adenylate cyclase
   forskolin-stimulated activity of, 98
   system, 207
Adolescents, caries and, 77
Adrenal
   dopamine, 152
   epinephrine, 152
   gland, 147–155
   glucocorticoid hormones, 169, 177, 178
   hyperplasia, 151, 154–155
   medulla, 147, 154
   medullary tumors, 54, 57, 147–148
   norepinephrine, 152
   weight, 151
Adrenocorticotropic hormone (ACTH), 183, 185, 197
Adult Oral Health Survey (NIDR1), 77
Adults, caries and, 77
Aedes aegypti, 242
Agarose, 128
Agglutinin, 130
Aging
   adenohypophysial cell morphology and, 191–200
   adrenal catecholamine changes in male Lobund–Wistar rats and, 147–156
   alterations in process of, 47
   biochemistry of, 259
   biological rate of, 135
   bone matrix changes and, 87–94
   compromise of male rat liver microsomal monooxygenases and, 283–288
   C-reactive protein and, 127–132
   degenerative changes associated with, 248
   dietary restriction and, 3–35, 97, 202
   effect of on serum hormone and blood chemistry changes, 135–146
   disease not synonymous with, 75
   endocrine systems and, 31–32
   environmental challenge and, 163
   enzymes and, 263, 277
   extracellular homeostatic and integrative mechanisms and, 102
   free radical damage and, 46, 101
   functional and biochemical parameters in, 229–239

312 / Index

α2u globulin effect on, 300–309
glutathione deficiency of, 242
immunological parameters and, 105–117, 178
lymphocyte subsets in germfree and conventional Lobund–Wistar rats and, 117–126
medullary structure changes in male Lobund–Wistar rats and, 147–156
metabolic parameters and, 265
mitogen concentrations and, 112
model of, 201–208
molecular changes in skeletal muscle and, 277–282
neural systems and, 31
normal, 75, 117
oral health in humans and animals and, 75–85
oxidative modification effect of on cellular protein turnover and, 269–276
pancreatic hormone changes during, 163–166
past and future research overview of, 7–25
peridontal disease and, 78
physiological process changes and, 27–28, 75
prevention of downside of, 25
primary processes of, 29
protein changes and, 259–267
psychological aspects of, 75
reality of, 3
regulation of process, 202
retardation of by dietary restriction, 97–103
small intestinal crypt cell production rate and, 217
sociological aspects of, 75
spontaneous diseases in Lobund–Wistar rats and, 51–60
underfeeding effect on, 27
via neuroendocrine system, 169–180
variability of change and, 1
Aging Model, 270
Alanine transferase, 136, 142
Albumin, 128, 136, 141–142
Aldolase, 260
Alfalfa meal, 41

Alkaline phosphatase, 136, 140–141, 213, 216, 219
Allogeneic
lymphocyte reactions, 106
spleen cells, 109–110
Alveolar bone loss, 79–80, 83
Alzheimer's disease, 8, 17
α-Amanitin, 293
Amidation, 261
Amines, 155
Amino acids
bone collagen, 89
C-reactive protein 127
C-terminal, 261
in enzyme residues, 269
MEM nonessential, 106–107
substitutions, 261
supplemental dietary, 41
see also specific amino acids
6-Aminonicotinamide, 206
Ammonium purpurate, 79
Amplification of error, 259
Amyloid, senile cardiac, 69–70
Androgens, 283, 302, 306
Anemia, 130
Anesthetics. See specific agents
ANOVA, 150, 235, 238
three-way, 106
Antibiotics, 193
Antibodies
anti-albumin, 128
anti-C3b, 128
anti-DNA, 117
anti-fibrinogen, 128
anti-transferrin, 128
auto-anti-idiotype, 125
monoclonal, 118–119, 123, 262, 295
anti-OX19, 120
biotinylated, 119
lymphocyte subsets and, 120
polyclonal
rabbit anti-rat cytochrome P-450, 285–286
rabbit anti-rat reductase, 285
see also Autoantibodies
Antibody mediated immunity, 106, 111
Antigen(s)
density, 125

environmental, 118, 123
lack of stimulus for, 140
stimulation of, 113
T cell, 125, 289–290
T-dependent, 112
Antioxidant defense system, cellular, 46, 247–256
Anti-P component, 70
Antisera, 183, 260
Anus, replicating tissues of, 212
Aortic strips, tension development by smooth muscle of, 28
Apoprotein B, 127
Araldite, 184
Argon laser, 119
Arthritis, 8
Ascites, fluid, 119
Ash, 41
  carcass content, 62–64
Aspartate transaminase, 136, 142
Assays
  colorimetric, 89
  DNA polymerase, 290
  DNA unwinding activity, 292–295
  insulin, 137–138
  Lowry, 128
  mitogen, 107
  primase, 290–291
  prolactin, 138
  protein, 271
  TBA, 101
  see also Immunoassays and Radioimmunoassays
Atherosclerosis, 248
Atomic absorption
  spectrophotometry, 89
  spectroscopy, 89
Atonia, intestinal, 39
ATPase, 277
  sarcoplasmic reticulum, 278–281
Atria, cardiac fibrosis and, 69–71
Atrophy
  oral mucosal, 75–76
  of oral-facial musculature, 76
  ovarian, 171
  pressure of kidney medullary regions, 56
  thymic at onset of sexual maturity, 117
  uterine, 171

Autoantibodies, 125; see also Antibodies
Avidin
  -biotin-peroxidase complex technique, 183, 185
  -phycoerythrin, 119

Baby boom, 13
Bacilli, tuberculosis, 10
Background subtraction, 119
Basal/induced activity hypothesis, 102
Basophil(s) 185
  adenomas, 196
B cells, 123
  change in number of with age, 117
  elimination of by nylon wool, 111
  polyclonal activators, 125
  spleen, 112
  T-dependent antigen responsive, 112
  see also B lymphocytes
Benign stromal hyperplasia, 55–57, 59
BHT, 41
Bile, 232
Bilirubin, 136, 142
Biochemical index, blood glutathione as of life span enhancement in diet-restricted Lobund–Wistar rat, 241–246
Biochemistry
  of aging, 259
  cellular, 259–309
  nutritional, 229–256
Bipyridyl reagents, 249
Birth
  life expectancy at, 21
  rates, 17
Bladder, dilated urine-filled, 56
Blood
  cell differential counts, 118
  chemistry
    dietary restriction effect on, 135–146
    parameters, 135–136, 139–144, 231
  glucose, 171
  glutathione (GSH) as biochemical index of life span enhancement in diet-restricted Lobund–Wistar rat, 241–246
  heparinized, 118, 249
  orbital, 106, 136
  peripheral, 105, 107, 118–119
  portal vein, 164

pressure, 98
   elevated, 8
   systolic, 28
urea nitrogen, 136, 142
vessels and cardiac fibrosis, 71
Blue agarose column chromatography, 295
B lymphocytes, 105, 124; see also B cells
Body
   composition analysis of Lobund–Wistar rats, 61–66
   fat, 204
   reduction of, 61, 202
   lean carcass mass, 63–65
   size, 61
   temperature, 171, 233
   weight, 139, 159, 203, 233
Bone
   calcium levels, 87–89, 92
   collagen, 89, 91
   density, 79
   mandibular, 82
   formation, 87, 94, 140
   Gla protein (BGP), 87–90, 92–93
   hydroxyproline, 91–92
   induction, 94
   loss
      alveolar, 79–80, 83
      mandibular, 83
   magnesium levels, 87–89, 91
   matrix
      age-related changes in organic and inorganic, 87–94
      inorganic, 87–94
      organic, 87–94
      parameters, 87–88, 93
   metabolism, osteoporosis and, 25
   osteoblastic mechanisms of laying down of, 25
   resorption, 87
   ultimate stress and, 88
Bouin's solution, 53
Bovine serum albumin (BSA), 137, 290
Brain
   dietary restriction effect on, 99
   lipofucsin, 231, 237–238
   metabolism, 204, 206
   phosphoglycerate kinase in, 260
   post-mitotic cells in, 212
   protein synthetic activity in, 98
   rat, 278
   recognition of altered energy balance, 202–203
   stem, 205
   uptake of metabolites and energy by, 204
Breast adenofibroma, 58
Brush border
   disaccharidases, 219
   enzymes, 213, 219
   microvillus, 213
Butanol, 232

Caenorhabditis elegans, 263
Calcification, 82
Calcitonin
   gene, 32
   plasma, 28
Calcium, 127, 136, 142
   bone, 87–89, 92
   ion, 277, 279
   phosphate, 41
Calories
   consumption of, 52
   restriction of, 29, 61, 65, 98, 169, 171, 177, 219, 283
   and effect on immune function in aging rats, 105–116
Cancer
   free radicals and, 248
   hematopoietic, 111
   lymphoid, 111
   mammary, 172–173
   risk, 66
   stomach, 10
   see also Neoplastic disease, Tumors, and *specific cancer types*
Cannibalism, among fullfed rats, 53
Carbohydrates, 98, 214, 301
Carbon dioxide, 136, 141–142
Carbonyl groups, 270–274
Carboxyglutamic acid protein, bone, 88–90, 92–93
Carcass
   ash content, 62–64
   fat mass, 64–65
   lean body mass, 63–65
   size data, 63
   water content, 62, 64

weight, 65
Carcinogenesis
    multistep theory of, 197
    pre-neoplastic stage of chemically
        induced, 217
Carcinoma, 54–55
Cardiac
    fibrosis in aging germfree and conventional Lobund–Wistar rat, 69–74
    output, 233
    of germfree animals, 39
    puncture, 118
Cardiomyopathy, 29, 69, 72–73
Caries
    cervical, 76
    coronal, 76, 79
    incidence studies of, 77
    rates of at different ages, 77
    recurrent, 77
    root, 76–77, 79
    secondary, 77
Carrier-facilitated absorption, 235
Casein, 47
Castration, 178
    cells, 195–196
Catalase, 101, 249, 251, 253–254
Catecholamines
    age-related changes in adrenal in male Lobund–Wistar rats and, 147–156
    hypothalamic, 170
    synthesis of, 148
Cathepsin D, 269, 273
C cells, thyroid, 32
Cecum
    enlarged, 39, 141
    Lobund–Wistar germfree rat, 42, 46
    twisted, 39
Cell(s)
    adrenal medullary, 57
    allogeneic target, 117
    antioxidant defense systems of, 248
    chromophobic, 185
    early passage, 263
    hematopoietic, 12
    hepatoma, 55
    inflammatory, 214
    late passage, 259–260, 263
    -mediated immunity, 105, 111

PAS negative, 185
PAS positive, 185–186
post-mitotic, 212
proliferative populations of, 102
prostate adenocarcinoma, 56
replication, 102, 212
serous, 80
tumor, 55–56
see also Cellular and *specific cell types*
Cellular
    antioxidant defense system, 46, 247–256
    biochemistry, 259–309
    energy costs, 100
    enzymes, 269, 273
    metabolism, 205
    protein turnover, 255
    and oxidative modification, 269–276
Centrifugation, 194, 231, 271
Central nervous system (CNS), 202
Cervical caries, 76
Chemotherapeutic agents. *See specific agents*
Chest pain, crushing, 10
Chickens, embryonic brain and, 294, 296
Children, caries and, 77
Chloride, 136, 141–142
Chloroform, 238
Cholesterol, 136, 140, 226
    plasma, 27
    serum, 144, 237
Chromaffin cells, 147
Chromatin, 101–102, 291–292
Chromatography
    blue agarose column, 295
    DE-32 column, 292, 295
    hydrophobic interaction column, 291, 295
    phosphorylcholine affinity, 128
Chronic toxicity studies, 47
Chymotrypsin, 262
Citric acid, 231
Colitis, ulcerative, 217
Collagen
    bone, 88–91
    deposition in heart tissues, 70–73
    hydroxyproline in, 91
    purified, 194
Collagenase, 194
Colon, 219
    epithelial cell turnover in, 212

Colorimetric assay, hydroxyproline determination and, 89
Columns
    DE-32, 292, 295
    Dowex-50W, 231
    octyl-sepharose, 291
    Sephadex, 137
Complement activators, 130
Conconavalin A, 106, 108–109, 113, 117
Congo red stain, 70
Conversion hysteria, 10
Corn
    cob bedding, 52
    ground, 41
    oil, 41, 172
Coronal caries, 76, 79
Coronary vessel disease, 72
Corticosteroids, 197
Corticosterone, 302
Corticotroph(s) 195
    hyperplasia, 196–197
    proliferation, 197
Corticotropic hormone releasing hormone (CRH), 188, 197
C-reactive protein, 127–129
    in aging Lobund–Wistar rats, 127–132
Creatinine, 136, 142
Crypt
    cells, 213, 215–219
    colonic, 217
    proliferation, 216
    -villus subjunction, 213
Cuts, 261
Cycloxygenase, 226
Cysteine, 242
Cytochrome
    c, 285
    P-450, 285–286
Cytoplasm, 186
    castration cell, 196
    enzymes, 264
    granules in, 79
    ground glass-appearing, 55
    hepatocytes, 54
Cytoskeleton, 125
Cytosol
    liver, 249
    proteases, 269, 273

Dark phase, light cycle, 31
DE-32 column chromatography, 292, 295
Deamidation, 261
Death
    age at of Lobund–Wistar rat, 62
    causes of in Lobund–Wistar germfree rat, 45
    measurement of time of, 21
    spontaneous, 29
Decalcification, 79
Decapitation, 148
Degradation, proteolytic, 272
Dehydrogenases, 272
Delayed hypersensitivity reaction, 117
Dementia, 17
    projected statistics for in U.S., 19
    see also specific conditions
Dentition, loss of, 75
Dephosphorylation, 261
Deproteinization, 148
Desensitization
    heterologous, 207
    homologous, 207
Deterioration
    immunologic, 61
    physiologic, 61
Detoxification, 99
Diabetes, 302
Dicalcium phosphate, 41
Diet, 51
    age-related changes in organic and inorganic bone matrix constituents and, 87–94
    dead microbes in, 112
    fibrous, 83
    as heart disease risk factor, 12
    influence of on rat lifespan, 47
    L-462, 234
    L-485 natural ingredient, 41, 53, 106
    mitogen concentrations and, 112
    RAtion Netherlands (RAN), 234
    semisynthetic, 28, 218
    soy-based, 73
    steam-sterilized, 53
    tryptophan deficiency in, 169
    see also Dietary restriction
Dietary restriction
    action of via endocrine system to produce effects, 171–175

adenohypophysial changes and, 181–190
age-related changes in rat small intestine
 and, 218–219
aging and, 3–35
blood glutathione and, 241–246
bone matrix parameters and, 88
brain metabolism changes and, 206
effects of on body composition and size in
 germfree and conventional
 Lobund–Wistar rats, 61–68
energy metabolism and, 99
food intake control and, 201–208
gastrointestinal cell growth and, 211–220
$\alpha 2u$ globulin effect on, 300–309
hypothalamic metabolism and, 201–208
lifespan enhancement and, 28–29, 78,
 234–238, 241–246, 248, 255, 303
lymphocyte subsets in germfree and conventional Lobund–Wistar rats and,
 117–126
mechanistic possibilities for actions of,
 30–33, 97–102
metabolic rate and, 30, 32, 234, 239
microsomal liver monooxygenases and,
 283–288
natural killer cell activity and, 102
neurological and immunologic interpretations and, 102
oral health in humans and animals and,
 75–85
overview of effects of, 27–35
pathologic lesions and, 28–29
plasma Vitamin E levels and, 254
-resistant parameters, 98
retardation of diseases and aging by,
 97–103
serum hormone and blood chemistry
 changes in aging Lobund–Wistar
 rats and, 135–146
and serum somatomedin-C/insulin like
 growth factor-1 in young, mature,
 and old rats, 157–162
systolic blood pressure and, 28
thyroid function and, 32
 see also Diet
Differentiation
 delay in, 216
 gastrointestinal cell, 213

hematopoietic cell, 112
lymphocyte, 112, 125
5α-Dihydrotestosterone, 302
Dihydroxybenzylamine (DHBA), 148–149
Dihydroxymandelic acid (DHMA), 149, 151
7,12-Dimethylbenz[a]anthracene, (DMBA),
 65, 172
2,4-Dinitrophenylhydrazine, 271
Disaccharidases, 213, 219; *see also specific
 enzymes*
Disease
 aging and risk for, 61, 75
 effect of on family and whole social network, 17
 infectious, 51, 129
 kidney, 144, 239
 liver, 144
 lymphoproliferative, 117
 metabolic, 277
 model systems of, 52
 neoplastic, 29, 47, 78, 111
 occurrence of in only portion of aging
  population, 75
 pathogenesis of, 30, 248
 patterns of, 10
 periodontal, 76–78
 prostate, 55–56
 renal, 78
 retardation of by dietary restriction,
  97–103
 spontaneous in aging Lobund–Wistar
  rats, 51–59
 systemic, 76
 see also specific diseases
DNA, 101–102, 293
 antibodies, 117
 calf thymus, 290, 292–293
 chain initiation stimulation by NPF1 from
  rat liver, 289–299
 crypt cell incorporation of
  [^{3}H]Tthymidine into, 213
 damage, 256, 259
 errors in, 265
 IMR-32, 293–295
 ligase, 289
 liver, 232, 236–237
 muscle, 232, 236
 nuclear, 216

polymerase, 264
  α, 264, 289–291, 293–296
  α2, 292
  β, 264
  holoenzyme, 295
  repair, 101–102, 264
  replication, 264, 289–290
  sequences, 290
  SV-40, 296
  synthesis, 292–293, 295
  transcription, 289, 295
  translation, 289
  unwinding activity assay, 292–295
DNAase, 293
Dogs, thymus function in aging, 178
Dopamine, 149–155, 187, 193, 206
  -β-hydroxylase, 147–148
Dowex-50W column, 231
Drosophila spp., 263
Dual electrochemical detection, blood glutathione and 245
Duncan's multiple range test, 89, 139, 150
Duodenum, 215, 217–219

EDTA, 128, 149, 271, 278
Efficiency, metabolic, 99–100
Eicosanoids, 226
Elderly, increase in numbers of, 7, 13
Electron microscopy, 147
  adenohypophysial studies and, 181–190
Electrophoresis, SDS-PAGE, 128, 292, 301, 305
ELISA immunoassay, 285, 286
Endocrine
  function, 135, 145, 147
  glands, 52
    benign adenomas of, 39
  lesions, 57, 147
  -regulated glands and pathological changes in, 52–53
  system and aging processes, 31–32
  see also Endocrinology
Endocrinology, 135–166; see also Endocrine
Endoplasmic reticulum, 196, 213
  rough, 185–186, 195, 197
Endpoint studies, 230
Energy
  balance regulation, 202–204
  coupling during ATP formation, 100

diet-induced, 100
efficiency of usage, 99–100
metabolism, 99–100, 103
restriction, 98, 255–256
thermic, 100
thermoregulatory, 100
work-induced, 100
Enolase, 260–262
Environment(al)
  age-related changes in organic and inorganic bone matrix constituents and, 87–94
  antigens, 118, 123
  stress and aging, 117, 163
Enzyme(s)
  adaptation to stimulation, 163
  aging and, 263, 277
  brush border, 213, 219
  cellular, 269, 273
  converting activity between epinephrine and norepinephrine, 154
  cytoplasmic, 264
  glycolytic, 277
  hepatocyte, 273
  homeostasis, 219
  intestinal, 215–216
  lysosomal, 264
  mammalian, 262
  mitochondrial, 264
  old-type, 260–261
  oxidative damage and, 269, 271
  proteolysis, 269, 274
  radical scavengers, 101
  structural studies of, 265
  testicular, 273
  transcriptional mistakes and, 264
  unaltered, 264
  young-type, 260–261
  xenobiotic metabolizers, 101
  see also Holoenzymes and *specific enzymes*
Eosin, 53, 79, 183
Epidemiological studies, periodontal disease trends and, 77
Epididymal fat pad, 43–46
Epinephrine, 147, 149–151, 153–155, 172, 233
  postreceptor processes and, 27–28

Epithelial cells, intestinal, 212
Epoxy resin, 194
Error Catastrophe Hypothesis, 259–260, 265
Erythrocyte(s), 249
 conglomerated, 185
 receptors, 125
Esophagus, epithelial cell turnover in, 212
Esters, phosphate, 127
Estradiol benzoate, 172–173
Estrogen, 173–174, 177, 193, 302
Estrous cycle, 171, 174, 176–177
Ethanol, 194, 291
Ethylene bromide, 232
Euthanasia, 4
Evisceration, 62
Exercise, as heart disease risk factor, 12
Exocytosis, 186
Exsanguination, 53, 62, 183, 193

Fat(s)
 body, 62–63, 98, 204
  life expectancy and, 30–31
  reduction of, 61, 202
 carcass mass, 63–65
 dietary, 41
  absorption, 226
  increased consumption of and heart attack rates, 11
  restriction of, 99
 epididymal pad, 43–46
 -free carcass weight, 63
 soluble vitamins, 254
Fatty acid(s)
 branched chain volatile, 226
 compositional changes of in germfree and conventional young and old rats, 221–226
 membrane 22:5, 33
 metabolism, 205
 polyunsaturated (PUFA), 226
 short chain volatile, 226
 synthesis, 205
 see also specific fatty acids
Feeding
 behavior
  control of, 203
  hypothalamic regulation of, 204
  following starvation, 213

Feminization, male rat liver microsomal monooxygenase and, 286
Femur, maximum breaking force for fracture of, 88
Ferric chloride, 249
Fetal calf serum (FCS), 106, 194
Fiber, 41
 dietary, 83
Fibrinogen, 128
Fischer 344 rat
 ad libitum fed, 214, 304–305
 adrenal gland, 147
 blood chemistry analysis and, 144
 body fat content of, 30–31
 cardiomyopathy and, 29, 73
 chronic nephropathy and, 28–30, 73
 crypt cellularity in, 215
 diet-restricted, 73, 304–305, 307
 food restriction studies and, 27–32
 $\alpha 2u$ globulin and, 304–305
 interstitial cell testicular tumors and, 283, 306
 intestinal
  enzyme activities in, 216
  epithelial cell proliferation in, 216–217
 male, 147, 283
 microsomal monooxygenases in, 285
 neoplastic disease in, 29
 pathogen-free, 27
 perfused livers, 32
 refeeding and, 215–217
 sacrifice of, 218
 senescent, 217
 starvation and, 215–217
 weight changes in, 214
Fisher Leukostain, 118
Flow cytometry, 118–119
Fluorescein, 119
 conjugated F(ab)'2 goat anti-mouse Ig, 119
 isothiocyanate (FITC), 120
Fluorescence
 green, 119
 red, 119
 two color, 119–120
Follicle stimulating hormone (FSH), 172, 174, 183, 185–188

Food restriction. See Dietary restriction
Formic acid, 88–89
Forskolin-stimulated adenylate cyclase activity, 98
Fractures
  femur, 88
  hip, 18, 24–25
Free radicals, 32, 46, 99–101, 169, 176–177, 254–256
  oxygen, 30, 46
  scavenger enzymes of, 101, 248
  superoxide, 249
  theory of, 259
Futile cycles, 100

Gamma-aminobutyric acid (GABA), 205–206
Gangliocytoma, 197
Gastric parietal cells and autoantibodies, 125
Gastrin, 233–234
Gastrocnemius muscle, weight of in Lobund-Wistar germfree rat, 43–46
Gastrointestinal
  cells
    cycling time, 214
    dietary restriction effect on growth of, 211–220
    migration of, 214
    production of, 213
    physiology, 211–226
    tract, 212, 214
Gels, two-dimensional, 263
General linear models procedure, 89
Gene(s)
  calcitonin, 32
  expression, 101–102
  $\alpha 2u$ globulin, 32, 303–305
  products, 289–290
  transcription, 304–305
Germfree status
  effects of, 232–234
  on immune function in aging rats, 105–116
  metabolism and, 256
Gerontology, dietary restriction mechanisms of action and, 103
Ghettos, age-specific mortality statistics and, 21
Gingival sulcus, hair impactions in, 83

Globulin, 136, 141
  $\alpha 2u$, 32, 283, 287, 300–307
  serum, 140
Glomerular sclerosis, 28
Glucagon, 165
  lipolytic action of, 27
Glucocorticoids, 31
  adrenal, 169, 177–178
Glucokinase, 163–164
Glucose, 31, 136, 138, 163–165, 204
  blood levels, 171
  carbon, 206
  metabolism, 205–206
  oxidation, 205
  serum, 139, 144
Glucose-6-phosphate, 271
  dehydrogenase, 270, 273–274
Glutamate decarboxylase, 205–206
Glutamine, 106, 271
  synthetase, 270, 273–274
γ-Glutamyl transferase, 136, 142, 271
Glutaraldehyde, 183, 194
Glutathione, 242–243, 245–246
  peroxidase, 101, 249, 252–253, 255
  reductase, 249
Glycation of hemoglobin, 31
Glyceraldehyde-3-phosphate dehydrogenase, 261
Glycerol, 290
  velocity gradient, 294
Glycine, 128
Glycosylation, 127
  end products, 31
Golgi apparatus, 186, 195–196
Gomori's reticulum stain, 70
Gonadal hormones, 177–178
Gonadectomy, 188
  basophil adenoma and, 196
  cells, 198
  pituitary adenoma and, 197–198
Gonadotroph(s), 186, 195–196
  adenoma, 183–185, 187–188, 196–198
  hormone releasing hormone (GnRH), 188
Gonadotropins
  effect of elevation on ovaries and uterus of underfed rats, 171–172
  immunoreactive, 198
  serum, 145

Gordon–Sweet method, 183
Gradients
  discontinous sucrose density, 291–292
  glycerol velocity, 294
Granules, cytoplasmic, 79
Growth
  cells, 12
  gastrointestinal cell, 211–220
  factor-1, 157–162
  hormone, 157, 177–178, 183, 197, 302
  immunoreactive, 185
  releasing hormone (GRH), 188, 197
  pattern of Lobund-Wistar germfree rats, 42
  serum, 170
  slowing rate and duration of, 61
Guanidine, 261
Guinea pigs, gonadotropic hormone effect on, 171

Hair impactions, in gingival sulcus, 83
Half-feeding, 173, 176
Haloperidol, 172–173
Halothane, 41, 53, 69, 106, 158, 183, 193
Hamsters, pentraxin and, 129
Hank's Balanced Salt Solution, 106
Heads, Lobund-Wistar rat, 79
Health, patterns of, 10
Hearing deficits, 8
Heart
  attack, 10
  collagen deposition in, 69–73
  disease, 8, 10, 12
  exsanguination of exposed, 53
  human, 72
  Lobund-Wistar germfree rat, 43–46
  rat, 72
  size of in germfree animals, 39
  weight, 232–233
  see also specific subunits of, 69
Heat
  production, 99
  sensitivity, 261
Helicase, 289
Hematopoietic cancers, 111
Hematoxylin, 53, 79, 183
Hemocytometry, 194
Hemoglobin, glycation of, 31
Hemorrhage, black measles and, 10

Hepatic
  microsomal monooxygenase, 284, 286–287
  parenchyma cells, 301
  proteins, 283
  see also Liver
Hepatocarcinoma, 55
Hepatocyte(s), 163, 237, 270–271, 273–274, 303
  cytoplasmic, 54
Hepatoma, 54–55, 130
HEPES buffer, 271
Heptanesulfonic acid, 149
High pressure liquid chromatography (HPLC), 149, 245, 262
Hip fractures, 17
  age-specific incidence of among white women, 24–25
  annual statistics for in U.S., 18
Histology
  adenohypophysial studies and, 181–190
Holoenzymes, 295; *see also* Enzyme(s)
Hormone(s), 135, 169
  adenohypophyseal, 185, 192
  administration and reversal of inhibitory effects of underfeeding, 172–174
  adrenal glucocorticoid, 169, 177–178
  estrogenic, 302
  gonadal, 177–178
  growth, 157, 177–178
  hypothalamic, 170
  pancreatic changes in during aging, 163–166
  pituitary, 145, 170–171, 255
  post-translational processing of, 145
  response status, 206
  serum, 142, 144, 170, 231, 233
    dietary restriction effect on, 135–146
  steroid, 302
  thyroid, 177–178
  see also specific hormones
Human(s)
  blood glutathione values, 242
  C-reactive protein, 127, 129
  disease as usual cause of death of in old age, 177
  gonadotroph adenoma, 197
  heart, 72

322 / Index

null cell adenoma, 198
oncocytic adenoma, 198
Humoral immunity, 105
Hydrazone, 271
Hydrogen
  chloride, 89
  peroxide, 249–250, 255
Hydroperoxide, 249
Hydrophobic interaction column (HIC) chromatography, 291
16α-Hydroxylase, 286
Hydroxyproline, 89, 91, 92
Hyperadrenocorticism, 31
Hyperchromatic cells, medullary lesions and, 57
Hyperlipidemia, 144
Hyperplasia
  adrenal, 147, 151, 154–155
  benign stromal, 55–59
  castration cell, 196
  corticotroph, 196–197
  ileal, 218
  lactotroph, 186–188, 193
  nodules and, 185
  prolactin cell, 193
  somatotroph, 197
Hyperprolactinemia, 182, 192
Hypertension, 57
Hypocorticism, 196
Hypogonadism, 188, 196
Hypophysectomy, 169, 171, 302
Hypothalamic
  catecholamines, 170
  dopamine concentration, 187
  fatty acid oxidation, 205
  hormones, 170
  lipid content, 204
  metabolism and dietary restriction, 201–208
  neurons, 174, 176
  prolactin inhibition, 192
  regulation of feeding behavior, 204
  regulatory peptides, 197
  see also Hypothalamus
Hypothalamus
  lateral, 204–205
  medial, 204–205
  ventromedial, 204, 206

see also Hypothalamic
Hypothyroidism, 144, 196
Hysterical paralysis, 10

^{125}I, 137
Ig, 125
  fluorescein-conjugated F(ab)′2 goat anti-mouse, 119
  mouse, 119
  rat, 119
  serum, 113, 117
IgA, 119, 125
IgG, 125
IgM, 125
Ileal hyperplasia, 218
Ileocecal valve, 218
Ileum, 217–218
Immune system, 130, 178
  aging rats and, 105–116
  alterations in as part of aging, 117
  deterioration of, 61
  dietary restriction and, 102
  underfeeding effects on, 177–178
  see also Immunity
Immunity
  antibody-mediated, 106
  cell-mediated, 105, 111
  humoral, 105
  see also Immune system
Immunoassays
  ELISA, 285–286
  see also Assays and Radioimmunoassays
Immunocytochemistry, adenohypophysial studies and, 181–190
Immunodiffusion techniques, 128
Immunoprecipitability, 286
Immunotitration, 261
Inbreeding, 52
Indigestion, misdiagnosed as heart attack, 11
Inflammatory cells, 130
Insulin, 98, 144, 163–165, 233, 302
  antilipolytic effect of, 98
  assay, 137–138
  -like growth factor-1, 157–162
  plasma, 28
  serum, 136–139, 142
Interferon (IFN), 112
Interleukin
  2 (IL2), 112, 117

3 (IL3), 112
Interstitial cells, testicular tumors and, 283
Intestinal
   atonia, 39
   enzymes, 215–216
   epithelial cells, 212
   see also Intestine
Intestine
   proximal, 215–216
   replicating tissues in, 212–213
   small, 213–214, 219
      bacterial degradation in, 235
      diet restriction and age-related changes in, 218–219
      histological changes in, 215
   see also Intestinal and *specific substructures*
Ionic balance, 141
Islets of Langerhans, 165
Isocitrate lyase, 260
Isoelectric focusing, 128, 261
Isolaters, 40–41
Isoniazide, 10

Jaws
   bone density in, 82
   bone loss in, 83
   calcification of, 82
   Lobund–Wistar rat, 79
Jejunum, 215–216, 218–219

KB cells, 295
Ketone bodies, 27
Kidney(s)
   α2u, 301
   cardiac fibrosis and, 69
   collagen deposition in, 71
   disease, 144, 239
      as major cause of death in most rat strains used in aging research, 47
   function, 141
   hypertension and, 57
   lesions, 28
   maltase, 260, 262
   pathology, 135
   pressure atrophy of medullary regions of, 56
Kinetics
   oral mucosal, 76
   skin, 76

L-462 diet, 234
L-485 natural ingredient diet, 41, 47, 52–53, 106, 158, 230
Lactase, 213, 215, 219
Lactotroph(s), 195
   adenoma, 182, 184–185, 187–188, 193, 195
   hyperplasia, 186–188, 193
Lanthanum oxide, 89
Laser, argon, 119
Lead citrate, 184, 194
Lesions
   endocrine, 57
   kidney, 28
   lung, 57
   medullary, 57, 59
   metastatic, 55
   palpation of, 47
   pancreatic, 57
   parathyroid, 57
   pathologic and food restriction, 28–29
   storage, 54
   thymus, 57
   see also Neoplastic disease, Tumors, and *specific types of lesions*
Leukemia, 130
Leukocyte(s), neutrophilic, 56
Leupeptin, 271
Life
   expectancy, 13
      at birth, 21
      body fat and, 30–31
      statistics for
         at age seventy five, 23
         at birth, 22
   extension, 8, 12, 46, 106, 145, 255
   last year or two of, 7–8
   quality of, 3–4
   span, 29
      dietary restriction and, 248
      germfree state and, 232

obesity and, 202
pathology and, 39–132
Light
  bodies, 186
  cycle dark phase, 31, 148
  effect of on FSH, 172
  green filtered ultraviolet, 70
  microscopy, 183, 194
  polarized white, 70
Limulus polyphemus, pentraxins in, 127
Linearity of physiological function loss associated with aging, 75
Lipid(s)
  hypothalamic content of, 204
  peroxides, 32–33, 100–101, 248
  plasma, 253
  serum, 144, 237
Lipofucsin, 100
  brain, 231, 237–238
Lipopolysaccharides, 106
  LPS, 107, 109–110, 113
  STM, 107, 109–110, 113
Lipoproteins, 130
  E-containing, 127
Lipoxygenase, 226
Liquid scintillation counting, 107
  multichannel, 290–291
Liver
  beta-oxidation in, 98
  cardiac fibrosis and, 69
  catalase activity, 251, 253
  cell size in, 231
  collagen deposition in, 71
  cytosol, 249
  disease, 144
  DNA, 232, 236–237
  function, 141
  glucose-stimulated insulin availability to, 164
  Lobund–Wistar germfree rat, 43–46
  membrane receptors, 130
  microsomal monooxygenases, 283–288
  mouse, 101
  NADPH-cytochrome reductase in, 261
  NPF-1 purification from, 291–295
  pathological changes of in diet-restricted rats, 53
  pentraxin production in, 127

per fused Fischer 344 rat, 32
phosphoglycerate kinase in, 260
post-mitotic cells in, 212
protein content, 236
rat, 283–299, 278
selenium-dependent glutathione peroxidase in, 253
stimulation of DNA chain initiation by NPF1 from, 289–299
superoxide dismutase activity, 251–252
tumors, 39, 54–55, 59
weight, 232–233
Lobund Aging Study, 40, 44, 270, 284
  design of, 39–49
Lobund–Wistar rat
  adenohypophyseal cell morphology in aging, 191–200
  ad libitum fed, 165, 306
  adrenal parameters in, 150
  adrenomedullary tumors in, 147
  age-related changes in adrenal catecholamine levels and medullary structure in male, 147–156
  aging and, 69–74, 181–239
  alveolar bone loss mean scores in, 80
  blood glutathione and, 241–246
  body composition and size in conventional and germfree, 61–68
  bone Gla protein content in tibial midshaft of, 90
  calcium content in tibial midshaft of, 92
  cardiac fibrosis in aging germfree and conventional, 69–74
  causes of death in, 45
  chronic nephropathy and, 47
  conventional, 39–40, 42–47, 52–53, 55–56, 61–190, 195, 230–231, 233–239, 243, 249–250, 252–254, 270, 272–275, 285
  C-reactive protein in aging, 127–132
  dietary restriction studies and longevity of, 39–48, 145, 165, 243, 249–253
  diet-restricted, 181–190, 195–196, 270, 272–275, 285, 305–306
  dopamine concentration in adrenal gland of, 152
  epinephrine concentration in adrenal gland of, 152
  evisceration of, 62

exsanguination of, 62
food-restricted, 181–190
fullfed, 52, 195–197, 243, 246, 249–253, 270, 272–275
functional and biochemical parameters in aging, 229–239
germfree, 52–56, 61–190, 195, 230–231, 233–239, 243, 249–250, 252–254, 256, 270, 272–275, 306–307
α2u globulin and, 305–306
gnotobiotic, 62, 136
growth of, 39–49
heads, 79
housing of, 106, 136, 157
hydroxyproline content in tibial midshaft of, 92
jaws, 79
life span enhancement and, 106, 241–246
liver, 291–292
magnesium content in tibial midshaft of, 91
male, 52, 106, 136, 139–144, 147–157, 183, 188, 230–231, 233, 307
  lean body weight in grams of, 43
  percent survival of at selected ages, 43
  spontaneous diseases in aging, 52–59
malignancy in, 111, 114
median survival age of, 46–47
microsomal monooxygenases in, 284–287
natural death of, 79
natural killer cell activity in, 114
nephrosis in, 144
norepinephrine concentration in adrenal gland of, 152
NPF-1 and, 291–292
nylon wool enriched spleen cells in, 111
pathology of, 106
peripheral blood of, 118
physiological and metabolic parameters of, 233
proliferative response of spleen cells from, 110
sacrifice of, 40, 42–43, 79, 89–90, 106, 136, 148, 157, 193, 249
serum bone Gla levels in, 93
serum hormone and blood chemistry changes in due to dietary restriction, 135–146

SMC/IGF 1 and, 157–160
spontaneous diseases in aging, 51–60
submandibular gland mean weight in, 82
survival of, 39–49, 244
T cell defects in, 113
tumors in, 58, 118
see also Wistar rat
Locomotor activity, spontaneous, 28
Long-Evans rat, adrenomedullary hyperplasia in, 147
Longevity, 3, 7, 45, 51, 59, 61
  blood glutathione and, 242–245
  factors contributing to, 66
  food restriction and, 28–29
  germfree state and, 78
  underfeeding and, 169, 303
Longitudinal Study of Aging, 76
Lowry assay, 128
Lung
  adenoma, 58
  cardiac fibrosis and, 69–70
  collagen deposition in, 71
  lesions, 57
  prostate adenocarcinoma cell spread to, 56
Luteinizing hormone (LH), 174, 183, 185–186
  serum, 170
Lymph nodes, 125
Lymphocytes, 108, 111, 118–119
  allogeneic reactions of, 106
  autoantibodies to, 125
  differentiation, 112, 125
  DNA polymerase in, 264
  fluorescent, 119
  immature, 125
  maps, 123
  membrane composition and fluidity, 105
  nylon wool-enriched, 106
  older, 125
  peripheral blood, 107
  subsets, 118–119, 124
    effect of dietary restriction and aging on in germfree and conventional Lobund–Wistar rats, 117–126
  syngeneic mixed reactions of, 106
  T-suppressor/cytotoxic (Ts/c), 122–123

young, 125
    see also B lymphocytes and T lymphocytes
Lymphoid
    cancers, 111
    tissues
        dietary restriction effect on, 99
        lymphocytes, 105
        macrophages in, 105
Lymphoma, YAC-1 cells, 107
Lymphoproliferative disease, 117
Lysine, 41
Lysosomes, 186, 194–195, 264

Macrophage(s), 105, 112
    activation state of, 106
    germfree, 113
    peritoneal, 113
    splenic, 113
    suppressors, 113
Magnesium
    bone, 87–89, 91
    chloride, 290–291
    thiazolidine carboxylate, 242
Malnutrition
    dietary restriction without, 105
    essential nutrient, 97
    undernutrition without, 27
Malondialdehyde (MDA), 32, 101, 231–232, 237–238
Maltase, 213, 215, 219, 260, 262
Mammary
    adenofibroma, 144
    cancer, 172–173
    tumors, 65, 172–173, 176–177
Mandibles. *See* Jaws
Mandibular molars, 79
Mapping, lymphocyte, 123
Masson trichrome stain, 70–72
Maturation, delaying of, 61
Maximum breaking force and femur fracture, 88
Measles, 10
    black, 10
Medicare, 12
Medullary
    lesions, 57, 59
    structure, age-related changes in male Lobund–Wistar rats and, 147–155

Membrane
    22:5 fatty acid, 33
    lipid peroxidizability, 33
    phospholipids, 237
Men
    life expectancy of at age seventy five, 23
    at birth, 22
    total population statistics for, 14–16
Meningioma, 186
2-Mercaptoethanol, 107, 278
Metabolic
    body rate per unit mass, 61
    disease, 277
    efficiency, 100, 103
    mass, 31
    pathways and translation of activity into neurochemical signal, 205–206
    see also Metabolism
Metabolism, 232
    bone and osteoporosis, 25
    brain, 204, 206
    cages, 231
    cellular, 206
    C-reactive protein, 130
    dietary restriction and, 234, 239
    fatty acid, 205
    germfree status and, 256
    glucose, 205–206
    hypothalamic and dietary restriction, 201–208
    neuronal, 206–207
    nitrogen, 157
    peripheral, 204
    see also Metabolic
Metaphase accumulation technique, vincristine-induced, 216–218
Metastasis, 55–56
    prostate adenocarcinoma, 53, 58–59
    tumor, 56
Methanol, 149, 238
dl-Methionine, 41
Methionine sulfoxide, 261
Methylation, 261
3-Methylhistidine (MEH), 231–233, 235–236
Mice
    autoimmune-prone, 117
    blood glutathione levels in, 242
    catalase and, 254

dietary restriction effects on, 97
disease as usual cause of death of in old age, 177
food restriction studies, 27–28
germfree, 39
glutathione peroxidase and, 255
Ig in, 119
immunocompetence in, 177
livers of, 101
serum tocopherol levels in, 252
spontaneous tumor formation in, 61
superoxide dismutase and, 254
thymus-dependent immunological function in, 177
see also individual species
Microflora, 40, 51–52, 106–114, 141–142
dead, in diet, 112
deleterious effects of on aging host, 39
effects of on oral health in humans and animals, 75–85
peridontal disease and, 78
Microsomal liver monooxygenases, age-dependent compromise of in male rat, 283–288
Microvilli, 213
Millonig's buffer, 194
Minerals, 29, 98; *see also specific minerals*
Mini-Mash cell harvester, 107
Mitochondria, 99, 196
autoantibodies to, 125
enzymes and, 264
studies of, 259
Mitogen(s)
assays, 107
pokeweed, 109, 112
T cell, 106, 108–109, 112
Mitomycin C, 107
Mitotic figures
adrenal medullary cell, 57
hepatoma cell, 55
Mixed function oxidases, 99
Models
aging, 270
blood glutathione, 242
conventional rat, 239
diet-restricted animal for the study of control of feeding behavior, 203

germfree, 230
oxygen toxicity, 270
periodontal research, 78
Molars, mandibular, 79
Monocytes, 123
Morbidity
body weight/height relationship and, 66
osteoporosis and, 87
Morphology, adenohypophysial cell, 191–200
Mortality
age-specific statistics, 21
body weight/height relationship and, 66
heart disease and, 12
osteoporosis and, 87
statistics for humans, 1900–1988, 9
Mosquitoes, 242; *see also individual species*
MRL/Mp-1pr mice, longevity and, 117
Mucicarmine, 79
Mucosa, oral
atrophy of, 75–76
kinetics of, 76
Muffle furnace, 63
Multifactorial analysis of variance, 150
Murexide, 79, 82
Muscle
dietary restriction effect on, 99
DNA, 232, 236
gastrocnemius weight of in Lobund-Wistar germfree rat, 43–46
glyceraldehyde check in, 261
protein, 231, 235–236
skeletal age-related molecular changes in, 277–282
smooth
aortic strip and tension development in, 28
autoantibodies to, 125
Mus spp., enzyme accuracy in, 264
Myocardial infarction, 10, 12
Myofibrils, 72, 235, 239
Myxoma, 130

NADP, 271
NADPH, 249, 285
Natural killer (NK) cells, 102, 109–111, 114, 118, 121–122, 124
splenic activity of, 106, 108
two color fluorescence and, 120
Nebenkerns, 185

Nematodes, 260-263; *see also individual species*
Neoplastic disease, 51, 78, 193
  dietary restriction and, 47
  in Lobund-Wistar rat, 111
  see also Cancer, Lesions, Tumors, and specific tumor types
Neovascularization, pituitary, 187, 193
Nephritis, 47
Nephropathy
  chronic, 28-30, 47, 51-52
  in Fischer rats, 73
  spontaneous, 58
Nephrosis, 144
Neurochemical signals, translation of metabolic pathway activity into, 205-206
Neuroendocrine system, 31, 102, 135, 145
  effects of on aging processes, 169-180
Neuroendocrinology, 169-208
Neuromodulators, 207
Neurons
  DNA polymerase β and, 264
  hypothalamic, 174, 176
  metabolism of, 206-207
Neuropeptide(s), 206
  Y (NPY), 207
Neurotransmitters, 206
Neutrophils, 107, 111
  peripheral blood, 108
Ninhydrin, 231
Nitrogen
  blood urea, 136, 142
  -free extract, 41
  metabolism, 157
Nodules, hyperplastic, 185
Nonhistone chromatin protein (NHCP), 292, 294-296
Norepinephrine, 147, 149, 150-155, 172, 206, 233
NPF1, 289-299
Nuclease, 292
Nucleic acids, 125; *see also specific nucleic acids*
Nucleotides, 301; *see also specific nucleotides*
Nursing homes, 8, 17
  projected number of residents of in U.S., 20

Nutritional biochemistry, 229-256
Nylon wool, and enriched spleen cells, 107, 109, 111
NZB/NZW mice, longevity and, 117
NZW-rabbits, C-reactive protein for immunization of, 128
Obesity
  cancer risk and, 66
  index of relative, 66
  life expectancy and, 206
Occlusal wear, 79
Octyl-sepharose hydrophobic interaction column chromatography, 291, 295
Oncocytic change, 198
Oncogenes, 188
Ophtalaldehyde, 231
Opsonin, 130
Oral
  cavity, 75-76
  facial musculature atrophy, 76
  health effects of aging, diet restriction and microflora on in humans and animals, 75-85
  mucosa, 76
Orbital blood, 106
Orcinol color reaction, 231
Ori-cores, 290
Osmium tetroxide, 194
Osteoporosis
  causes of, 87
  elderly women and, 87
  hip fractures and, 25
Ouchterlony technique, 128
Ovariectomy, 174
Ovaries
  atrophy of, 171
  corpora lutea in, 172
  function of inhibited by underfeeding, 176-177
  gonadotropin elevation and, 171-172
  weight of, 171
Overfeeding, decreased life expectancy and, 202
OX4 cells, 123-124
OX6 cells, 123-124
OX8 cells, 118, 120, 122, 124
OX19 cells, 120, 122, 124

Oxidases, mixed function, 99, 265, 269, 272, 274
Oxidative
 damage, 248, 259, 269
 products in urine, 98, 101
 enzyme activation, 269, 271
 modification and role of in cellular protein turnover and aging, 269–276
Oxygen, 270–274
 consumption by Lobund–Wistar germfree rats, 39, 46
 detoxification, 99
 free radicals, 30, 46
 metabolically generated active, 99–100
 molecular, 99
 toxicity model, 270
 use, 233

PA-3 cells, 295–296
Pain, radiating down left arm, 10
Palpation, abdominal, 53
Pancreas
 hormone changes in during aging, 163–166
 hypertension and, 57
 lesions of, 57, 130
 see also specific substructures
Paraffin, 69, 79, 183
Paralysis, hysterical, 10
Parathyroid
 adenoma, 57–58
 hormone, 28
 tumor, 130
Parenchyma, 55
Parietal cells, autoantibodies to gastric, 125
Parotid fluid, 76
PAS technique, 183
Pathology, life span and, 39–132
Pearls, 56
Pentameric discs, 127
Pentose, 205–206
Pentraxins, 127, 129
Pepstatin, 271
Peptides, 160, 262
 anabolic, 157
 hypothalamic regulatory, 197
 see also specific peptides
Perchloric acid, 148

Perfusion method, 270
Periodic acid-Schiff, 79
Periodontal disease, 76–78
 models of research on, 78
 pathogenesis of, 78
Peripheral metabolism, 204–205
Permissive cell factors, 289
Peromyscus spp., enzyme accuracy in, 264
Peroxidases, 101
Peroxidation
 damaging effects of, 231
 lipid, 32, 100–101, 248
Phenobarbital, 284
Phenylmethylsulfonyl fluoride (PMSF), 271
Phloxine-tartrazine, 79
Phosphate, 127, 137, 249
3-Phosphoglyceraldehyde dehydrogenase, 262
Phosphoglycerate kinase, 260–262, 277–279
Phospholipids, 280
 membrane, 237
 plasma, 27
Phosphorus, 136, 140–141
Phosphorylation, 261
Phosphorylcholine (PC), 127–128
Phycoerythrin, 119
Phylogenetic independence, 97
Physical labor, reduced and increased heart attack rates, 11
Physiology, gastrointestinal, 211–226
Phytohemagglutinin (PHA), 106–109, 113, 117
Pierce BCA Protein Assay method, 271
Pituitary
 adenoma, 182–186, 192, 197–198
 cells, 188
 function, 145
 hormones, 145, 170–171, 255
 in vitro studies of, 195
 neovascularization, 187, 193
 posterior, 184
 tumors 130, 182
Plaques, endothelial, 71
Plasma, 249
 calcitonin, 28
 cholesterol, 27
 glucose, 31
 insulin, 28

ketone bodies, 27
lipids, 253
membrane, 206
parathyroid hormone, 28
phospholipids, 27
triglycerides, 27
Vitamin E, 249-251
Pneumonia, 52
Pokeweed mitogen, 109, 112
Poly(A), 302
Poly(dC), 292
Polyethylene glycol, 291-292
Polypharmacy, 76
Polyunsaturated fatty acids (PUFA), 226
Postmenopause, 76
Postreceptor processes, 206
   epinephrine and, 27-28
Potassium, 136, 141-142
   chloride, 290-292, 294-295
Precipitin, 130
Pregnancies, teenage, age-specific mortality statistics and, 21
Prescription medications, xerostomic side effects of, 76-77
Primase, 289-290, 293-296
   assay, 290-291
   -initiated DNA synthesis, 292
   isolation of free by HIC, 291
Proestrus, 171
Progesterone, 174
Proinsulin, 137
Prolactin, 136, 172, 174, 177, 182, 187, 233
   adenoma, 145, 182, 184-185, 187-188, 192
   assay, 138
   cell hyperplasia, 193
   hypothalamic inhibition of, 192
   immunoreactive, 185
   inhibiting factor, 193
   serum, 138, 143, 145, 170, 173
Prostaglandin(s)
   dilator, 226
   E2, 113
   inhibition, 105
   PGI_2, 226
   vasoconstrictor, 226
Prostate
   adenocarcinoma, 53, 55-56, 58-59, 130
   disease, 55-56
   infections, 47
   tumors, 47, 53, 144
Prostatitis, 45, 55
Protease, 261-262, 274
   cytosolic, 269, 273
Protein(s), 41, 62, 98, 101, 136, 141, 250, 253
   abnormal, 259, 263-264
   age-related changes in, 259-267
   androgen-dependent, 302
   carbonyl content, 270
   carboxyglutamic acid, 87
   cellular oxidative modification and, 269-276
   C-reactive in aging Lobund-Wistar rats, 127-132
   factors
      NPF1, 289-299
      glomerular sclerosis and, 28
   hamster female, 129
   kinases, 206, 272
   liver, 236, 283
   low molecular weight, 289
   muscle, 231, 235-236
   myofibrillar, 235, 239
   nonhistone chromatin, 292, 294-296
   nuclear, 289
   oxidative damage to, 46, 269-275
   restriction of, 29
   serum, 127
   soy, 28, 47, 73
   synthesis, 32, 101-102, 145, 171, 254, 259, 263-264
   synthetic activity of in brain, 98
   turnover, 32, 102, 262-263
   urinary, 301
Proteinuria, 303-304, 306
Proteolysis, 269, 274
Pseudohypophysectomy, 171
Puberty, thymus size increase at, 178
Pyruvate, 106

Rabbit(s) anti-rat cytochrome P-450 polyclonal antibody, 285-286
   anti-rat reductase polyclonal antibody, 285
   C-reactive protein, 127
   gonadotropic hormone release in, 172

Radiation, 230
Radioimmunoassays
  antibody coated-tube, 137
  SMC/IGF one concentration, 158
  see also Assays and Immunoassays
RAtion Netherlands diet (RAN), 234
Rat(s)
  ad libitum fed, 32, 40, 46–48, 52–54, 62, 65, 78, 106, 111, 128, 145, 148, 158–159, 169, 171, 174, 178, 184, 203, 214, 218–219, 230, 255, 287, 304–306
  blood glutathione values and, 242
  B lymphocytes in, 124
  bone formation in, 87
  bone magnesium levels and, 91
  brain, 278
  catalase and, 254
  characteristics of experiental animals whose pituitaries were studied in vitro, 195
  conventional, 108, 111–112, 120, 123, 125, 221–226, 230
  C-reactive protein and, 130
  defective, 51
  degree of fibrosis in, 70
  diet-restricted, 27–28, 71–73, 79, 90, 94, 97, 107–109, 112, 120, 124, 129, 138–143, 145, 148–149, 154–155, 159, 182, 202–204, 218–219, 235–238, 304–305
  disease as usual cause of death of in old age, 177
  estrous cycle, 171
  fatty acid compositional changes in germ-free and conventional young and old, 221–226
  female, 171, 174, 302
  fullfed, 51, 53–55, 57–58, 62–64, 69, 71–72, 79, 83, 90, 94, 107–109, 111, 120, 123–124, 129, 138–144, 148–149, 171–172, 176, 235–238
  germfree, 39, 45–46, 59, 107–109, 111–113, 120, 123–125, 182, 221–226, 230–231, 242
  gnotobiotic, 139–142, 231
  gonadotropin elevation effect on ovaries and uterus of underfed, 171–172
  half-fed, 173, 176
  heart, 72
  hepatocytes, 270–271, 273–274
  histological changes in small intestine of, 215
  hormones in serum of young adult, 233
  hungry, 204
  Ig, 119
  immune function in aging, 105–116
  intestinal
    enzyme activities in, 215–216
    epithelial cell proliferation in, 216–217
  liver, 278, 283–299
  male, 40, 170, 178, 283–288, 301–303
  mature, 157–162
  muscle, 278
  natural killer cells in, 121
  NPF-1 purification from liver of, 291–295
  old, 157–226
  protein synthesis in livers of old, 259
  satiated, 204
  small intestine and age-related effects of dietary restriction on, 218–219
  superoxide dismutase and, 254
  Tc/s cells in, 123
  testes, 271–274
  thymus function in aging, 178
  thyroxin secretion in aging, 178
  T lymphocytes in, 121
  underfed, 174
  weanling, 53, 59
  young, 157–226
  see also individual species
Receptors
  abnormal, 188
  alteration of, 206
  beta, 28
  degradation of, 207
  glucagon influence on, 27
  liver membrane, 130
  recycling of to plasma membrane, 206
  sheep erythrocyte,, 125
Reciprocal skin transplants, 52
Recurrent caries, 77
Reductase, 284
Refeeding, 170, 174, 215–216

Renal
    disease, 78
    failure, 28
    see also Kidney(s)
Reticulin, 70, 183, 185
Rhesus monkeys, monooxygenases not apparent in outbred, 284
Rheumatic fever, 10
RIA diluent, 88–89
Ribonucleotides, 293; see also specific nucleotides
Risk
    assessments of drugs and chemicals, 52
    cancer, 66
    for diseases of aging, 61
RNA
    chain-initiated DNA synthesis, 290
    $\alpha 2u$ globulin, 302–304
    messenger (mRNA)
        SMC/IGF 1, 160
    polymerase, 264
        II, 293
RNAase, 232, 289
Root caries, 76–77, 79
RPMI 1640 medium, 106–107

S1 nuclease, 292
Saline, 128
Saliva, 76
Salivary gland
    ducts, 80–81
    weight, 79–83
Salmonella typhimurium, 107
Salmonella typhosa, 107
Sarcoplasmic reticulum, 278–281
Satiety, 202
Scarlet fever, 10
Scarring, heat valve, 71
Secondary caries, 77
Secretory granules, 185–186, 194–196
Selenium, 249
    -dependent glutathione peroxidase, 253, 255
Sella, tumor of, 186
Semisynthetic diet, 28
Senescence, 217, 303
Senile cardiac amyloid, 69–70
Sensory function, 99
    altered, 76

Sephadex column, 137
Serum
    bone Gla protein, 93
    cholesterol, 144, 237
    C-reactive protein, 128–130
    FSH, 170
    globulin, 140
    glucose, 139, 144
    gonadotropin, 145
    hormones, 170, 231, 233
        dietary restriction effect on, 135–146
        growth, 170
    Ig, 113, 117
    IgA, 125
    IgG, 125
    insulin, 136–139, 142
    lipids, 144, 237
    MDA, 231, 238
    mouse, 119
    normal rat, 128
    prolactin, 136, 138, 143, 145, 170, 173
    protein, 127
    SMC/IGF 1, 158, 160
    somatomedin-C, 157–162
    T3, 136–137, 143–145
    T4, 136–137, 143
    testosterone, 57, 136, 143, 145
    tocopherol, 252, 253
    triglycerides, 144, 237
    TSH, 136, 138, 143, 145, 170
Set point, 202
Sheep erythrocyte receptor, 125
Shunt pathways, 205–206
Single factor analysis of variance, 89
SJK-132-20, 296
SJK-237-71, 296
Skeletal muscle
    age-related molecular changes in, 277–282
Skin
    hemorrhage into in black measles, 10
    kinetics of, 76
    reciprocal transplants, 52
Smell, altered, 76
Smoking, as heart disease risk factor, 12
Social Security, 12
Sodium, 136, 142
    chloride, 41, 88, 271

hydroxide, 88
metabisulfite, 148
phosphate, 149, 278
Somatomedin-C
/insulin-like growth factor 1, 157–160
serum, 157–162
Somatostatin, 165
Somatotrophs, 197
Sonication, 271
Sorensen's buffer, 194
Sore throat, streptococcal, 10
Soy protein, 28, 41, 47, 73
Species variation, 127
Specific metabolic process, 61
Spectrofluorometry, 232
Spectrophotometry, 231
Spectroscopy
atomic absorption, 89
Spleen, 125, 177
B cell preparations, 112
cells, 107
allogeneic, 109–110
mitogen assays and, 106–107
natural killer, 106, 108
nylon wool-enriched, 107, 109, 111
proliferative response of, 109
syngeneic, 109–110, 112
size, 108, 111
T cell preparations, 112
weight, 108
Splenocytes, 106
natural killer activity of, 109–110, 111
Sprague–Dawley rat
female, 107, 163
male, 163
pancreatic hormones during aging and, 163–165
pituitary gonadotroph adenoma in, 183
spleen cells, 107
serum, 128
SR-ATPase from skeletal muscles of, 279
Staphylococcus aureus, 262
Starvation, 57, 170, 213, 215–216
Steam-sterilized diet, 53
Steroids, 302
Stimulator cells, 112
Stomach
cancer of, 10

epithelial cell turnover in, 212
Storage lesions, 54
Streptococcus spp., 10
Stroke, 8
Student-t-test, 70
Subtilisin, 269, 273
Sucrase, 213, 219
-isomaltase, 215
Sucrose, 291
density gradient, 291–292
postmenopausal females and, 76
Sulfation, 261
Sulfhydryl groups, 265
Superoxide
dismutase (SOD), 101, 249, 251–252, 254
radicals, 249
Suppressor cells, 105
Survival distribution function, 42
Syngeneic
mixed lymphocyte reactions, 106
spleen cells, 109–110, 112
Synthetases, 272

Taste, altered, 76
TBA assay, 101
T cell(s)
antigens, 125
cytotoxic, 112, 118
decline in function of with age, 117
defects, 112–113
-dependent functions, 125
see also T lymphocytes
Tension development, by smooth muscle of aortic strips, 28
Testes, 145, 271–274
Testicular
enzymes, 273
interstitial cell tumors, 283, 306
see also Testes
Testosterone, 136, 178, 233, 302
serum, 57, 143, 145
Teton method, 243–244
3,3-Tetraethoxypropane, 232
Thermal stability, 278
Thermic energy, 100
Thermogenesis, 100
Thiobarbituric acid, 232
Three way analysis of variance, 250–251

Thymidine, 107, 213, 216–217
Thymocytes mature, 125
Thymoma, 58
Thymus
   atrophy of at onset of sexual maturity, 117
   castration and, 178
   -dependent immunologic function, 177
   growth and function of, 177
   lesions of, 57
   lymphocyte differentiation and, 125
   size, 108, 111, 178
   testosterone and, 178
Thyroglobulin, 125
Thyroid
   C cells, 32
   function
      changes, 144
      and food restriction, 32
   hormones, 177–178
   stimulating hormone (TSH), 98, 183–185, 187
   serum, 136, 138, 143, 145, 170
Thyrotrophs, 184, 195–196
Thyrotropin, 98
Thyroxin (T4), 178, 289–290, 302
   serum, 136–137, 144–145
Tibia, 88–89
   bone Gla content of, 90
   calcium content of midshaft, 92
   magnesium content of midshaft, 91
Tissue culture studies
   adenohypophyseal cell morphology and, 191–200
   gonadotrophs and, 196
   pituitary and, 193–194
   supernatant fluid, 119
T lymphocytes, 105, 120–121, 124–125
   helper, 12, 118, 122, 124
   -independent functions, 125
   mitogens, 106, 108–109, 112
   pan-, 118, 120
   proliferative capacity of, 117
   spleen, 112
   suppressor, 118
   see also T cells
Tocopherol, 252–253; *see also* Vitamin E
Toluene scintillation system, 292
Toluidine blue, 184

Topoisomerase, 289
Total carcass fat, 63
Touch, altered, 76
TPM buffer, 292
Transferrin, 128
Transplants, reciprocal skin, 52
Trauma, lethal, 230
TR-cells, nonoverlapping populations of, 122
Trichloroacetic acid (TCA), 290, 292
Triglycerides, 136, 140
   hepatic, 144
   plasma, 27
   serum, 144, 237
Triiodothyronine (T3), 234
   serum, 136–137, 143–145
Tris, 290–292
Triton X-100, 280, 291
Trituration, 194
Trypan blue exclusion, 107, 194
Trypsin, 194, 262, 269
Tryptophan, 169
Ts/C lymphocytes, 122–123
Tuberculin test, 10
Tuberculosis, 10–11
Tumor(s)
   adrenomedullary, 54, 57, 147–148
   anthracene-induced mammary, 65
   benign, 55, 57, 147
   cells, 55–56
   endocrine, 147
   growth enhancement and, 197
   interstitial cell testicular, 283, 306
   liver, 39, 54–55, 59
   Lobund–Wistar rats and 42, 58, 110
   mammary, 172–173, 176–177
   metastatic, 56
   monohormonal, 192
   pancreatic, 130
   parathyroid, 130
   pituitary, 130, 182
   plurihormonal, 182, 192
   prostate, 47, 53, 144
   sella region, 186
   spontaneous formation of in mice, 61
   see also Lesions and *specific tumor types*
Turbatrix aceti, 260–261, 263
Two way analysis of variance procedure, 158

Ulcerative colitis, 217

Ultimate stress, bone, 88
Ultrastructure studies
   adenohypophysial cell morphology and, 191–200
   chromaffin cell, 147
   pituitary adenoma, 182, 186
Ultraviolet
   green filtered light, 70
   -induced DNA repair, 102
Underfeeding
   effect of on aging processes, 169–180, 303
   immune function and, 178
   ovarian function inhibition and, 176–177
   reversal of inhibitory effects of by hormone administration, 172–174
   severe, 171
   short-term followed by ad libitum feeding and temporary reproductive function improvement, 174–175
Undernutrition, without malnutrition, 27
Up-regulation antioxidant defense system, 256
Uranyl acetate, 184, 194
Urbanization, 8
Uric acid, 136, 141–142
Urine
   -filled dilated bladder, 56
   $\alpha 2u$ globulin and, 301–307
   3-methylhistidine in, 233, 236
   oxidative damage products in, 98
   xylose in, 231, 235
Uterus
   atrophy of, 171
   gonadotropin elevation and, 171–172

Vasoconstrictors, 226
Ventricles
   cardiac fibrosis and, 69–71
   collagen deposition in, 70
Villi, 213–214, 216, 218
Vincristine-induced metaphase accumulation technique, 216–218
Viruses, 40
Visual impairment, 8
Vitamin E, 254–255
   as free radical scavenger, 248
   plasma, 249–251
   see also Tocopherol
Vitamins, 41, 98; *see also specific vitamins*

Water, carcass content, 62, 64
Weight
   absolute, 43
   adrenal gland, 151
   average, 42
   body, 139, 159, 203, 233
      Lobund-Wistar germfree rat, 43–46
   carcass, 63, 65
   gain in Fischer 344 rat, 214
   of gastrocnemius muscle in Lobund-Wistar germfree rat, 43–46
   heart, 232–233
   liver, 232–233
   loss in Fischer 344 rat, 214
   ovarian, 171
   salivary gland, 79–83
   spleen, 108
White cells, 118
   differential counts, 106
   peripheral, 106–108
Wistar rat
   ad libitum fed, 230
   DNA in liver of adult male, 236
   see also Lobund–Wistar rat
Women
   age-specific hip fracture incidence among white, 24–25
   life expectancy of
      at age seventy five, 23
      at birth, 22
   osteoporosis in elderly, 87
   postmenopausal, 76
   total population statistics for, 14–16
World Health Organization, age-specific mortality statistics and, 21
W3/13 cells, 118, 120, 125
W3/25 cells, 118, 122, 125

Xanthine oxidase, 101
Xerostomic side effects, 76–77
Xylose, 214, 231, 235

YAC-1 mouse, lymphoma cells, 107
Yellow fever mosquito, 242